Series Editor

Prof. Dr. Michael J. Parnham
PLIVA
Research Institute
Prilaz baruna Filipovica 25
10000 Zagreb
Croatia

Inducible Enzymes in the Inflammatory Response

Derek A. Willoughby
Annette Tomlinson

Editors

Springer Basel AG

Editors

Prof. Derek A. Willoughby
Dr. Annette Tomlinson
Department of Experimental Pathology
St. Bartholomew's and
The Royal London School of Medicine and Dentistry
Charterhouse Square
London, EC1M 6BQ
UK

A CIP catalogue record for this book is available from the Library of Congress, Washington D.C., USA

Deutsche Bibliothek Cataloging-in-Publication Data
Inducible enzymes in the inflammatory response / ed. by D.A. Willoughby, A. Tomlinson - Basel ; Boston ;
Berlin : Birkhäuser, 1999
(Progress in inflammation research)
ISBN 978-3-0348-9755-6   ISBN 978-3-0348-8747-2 (eBook)
DOI 10.1007/978-3-0348-8747-2

© 1999 Springer Basel AG
Originally published by Birkhäuser Verlag in 1999
Softcover reprint of the hardcover 1st edition 1999
Printed on acid-free paper produced from chlorine-free pulp. TCF ∞
Cover design: Markus Etterich, Basel

ISBN 978-3-0348-9755-6

9 8 7 6 5 4 3 2 1

Contents

List of contributors

David R. Blake, The Royal National Hospital for Rheumatic Diseases, Upper Borough Walls, Bath BA1 1RL; e-mail: D.R.A.Blake@bath.ac.uk

Lee D.K. Buttery, Department of Histochemistry, Imperial College School of Medicine, Hammersmith Campus, Du Cane Road, London W12 0NN, UK; e-mail: l.buttery@rpms.ac.uk

Paul R. Colville-Nash, Experimental Pathology, William Harvey Research Institute, St. Bartholomew's and The Royal London School of Medicine and Dentistry, Charterhouse Square, London EC1M 6BQ, UK; e-mail: P.R.Colville-Nash@mds.qmw.ac.uk

Fernando Q. Cunha, Department of Pharmacology, Faculty of Medicine Ribeirão Preto, University of São Paulo, Avenida Bandeirantes, 3900, Ribeirão Preto, São Paulo 14 051-900 Brazil; e-mail: fqcunha@fmrp.usp.br

Manuel Distel, Department of Medicine, Boehringer Ingelheim International, 300 Beach Road, #14-02/04 The Concourse, Singapore, 199555

Sergio H. Ferreira, Department of Pharmacology, Faculty of Medicine Ribeirão Preto, University of São Paulo, Avenida Bandeirantes, 3900, Ribeirão Preto, São Paulo 14 051-900 Brazil; e-mail: shferrei@fmrp.usp.br

Derek Gilroy, Experimental Pathology, William Harvey Research Institute, St. Bartholomew's and The Royal London School of Medicine and Dentistry, Charterhouse Square, London EC1M 6BQ, UK; e-mail: D.W.Gilroy@mds.qmw.ac.uk

Hans-Peter Hartung, Department of Neurology, Karl-Franzens-Universität, Auenbruggerplatz 22, A-8036 Graz, Austria; e-mail: hp.hartung@kfunigraz.ac.at

Adrian J. Hobbs, Wolfson Institute for Biomedical Research, University College London, Rayne Institute, 5 University Street, London WC1E 6JJ, UK

Stephen Hyslop, Department of Pharmacology, Faculty of Medical Science, University of Campinas, Campinas, São Paulo, Brazil

Bernd Kieseier, Department of Neurology, Karl-Franzens-Universität, Auenbruggerplatz 22, A-8036 Graz, Austria; e-mail: bc.kieseier@kfunigraz.ac.at

Salvador Moncada, Wolfson Institute for Biomedical Research, University College London, Rayne Institute, 5 University Street, London WC1E 6JJ, UK

Michel Pairet, Department of Pulmonary Research, Boehringer Ingelheim, Birkendorfer Str. 65, D-88397 Biberach, Germany; e-mail: pairet@ing.boehringer-ingelheim.com

Mark Paul-Clark, Experimental Pathology, William Harvey Research Institute, St. Bartholomew's and The Royal London School of Medicine and Dentistry, Charterhouse Square, London EC1M 6BQ, UK; e-mail: M.J.Paul-Clark@mds.qmw.ac.uk

Julia M. Polak, Department of Histochemistry, Imperial College School of Medicine, Hammersmith Campus, Du Cane Road, London W12 0NN, UK; e-mail: j.polak@cpms.ac.uk

Michael P. Seed, Paneutics, P.O.Box 1358, Swindon SN3 4GP, UK; e-mail: m.p.seed@mds.qmw.ac.uk

Annette Tomlinson, Experimental Pathology, William Harvey Research Institute, St. Bartholomew's and The Royal London School of Medicine and Dentistry, Charterhouse Square, London EC1M 6BQ, UK; e-mail: A.Tomlinson@mds.qmw.ac.uk

Joanne van Ryn, Department of Pulmonary Research, Boehringer Ingelheim, Birkendorfer Str. 65, D-88397 Biberach, Germany; e-mail: joanne.vanryn@bc.boehringer-inhelheim.com

Dean Willis, Experimental Pathology, William Harvey Research Institute, St. Bartholomew's and The Royal London School of Medicine and Dentistry, Charterhouse Square, London EC1M 6BQ, UK; e-mail: D.Willis@mds.qmw.ac.uk

Derek A. Willoughby, Experimental Pathology, William Harvey Research Institute, St. Bartholomew's and The Royal London School of Medicine and Dentistry, Charterhouse Square, London EC1M 6BQ, UK; e-mail: D.A.Willoughby@mds.qmw.ac.uk

Vivienne R. Winrow, Department of Postgraduate Medicine, University of Bath, Claverton Down, Bath BA2 7AY, UK; e-mail: V.R.Winrow@bath.ac.uk

Preface

This volume will be of great value to all those researchers in the area of the inflammatory response, notably academics, clinicians and members of the pharmaceutical industry.

The book has in the main been restricted to three inducible enzymes, namely nitric oxide synthase (iNOS), cyclooxygenase (COX-2) and hemeoxygenase (HO-1), although matrix metalloproteinases, xanthine oxidoreductase and tissue transglutaminases are reviewed. The modulation of these enzymes is viewed as possible novel therapeutic advances in the area of inflammation and also cancer. The latter topic may well be the subject of a further book.

It will be interesting to observe the progress of such new therapies in the next decade. Already some of these enzyme modulators have been approved for the treatment of inflammatory disease, as evidenced by the new families of COX-2 inhibitors. We believe such advances will herald a series of new and exciting agents to be included in the clinician's armamentarium in the constant struggle against inflammatory disease.

The editors wish to thank all contributors to this volume on inducible enzymes. It should however be stressed that the views expressed by the authors are personal and do not necessarily reflect those of the editors. Indeed, the reader may find conflicting statements in a number of the chapters. We believe that this is entirely appropriate as this volume reflects the latest work in a rapidly developing area.

Finally, the editors would like to express their thanks to Birkhäuser and their helpful staff, in particular Janine Kern. We hope the reader enjoys the book and finds some thought-provoking issues in this wide spectrum of recent advances in a new and exciting field.

Derek Willoughby
Annette Tomlinson

Overview of COX-2 in inflammation: from the biology to the clinic

Michel Pairet, Joanne van Ryn and Manuel Distel*

Department of Biological Research and Department of Medicine*, Boehringer Ingelheim, Birkendorfer Str. 65, D-88397 Biberach/Riss, Germany

Discovery of COX-2: Key findings

Cyclooxygenase (COX) is part of a bifunctional enzyme, prostaglandin H synthase (PGHS), which catalyzes the first step in the conversion of membrane phospholipids (principally, arachidonic acid) into prostanoids (prostaglandins and thromboxanes). Prostanoids contribute to diverse physiological and pathological processes including inflammation. Non-steroidal anti-inflammatory drugs (NSAIDs), the prototype of which is aspirin, owe their therapeutic effects to blockade of prostaglandin synthesis through COX blockade [1]. NSAIDs are a heterogenous group of compounds, often chemically unrelated, which nevertheless share therapeutic effects (i.e. anti-inflammatory, analgesic and antipyretic) and side effects (i.e. gastrointestinal erosions, decreased renal function, inhibition of platelet aggregation). Until recently, it had been widely accepted that a common mechanism (COX inhibition) is responsible for both the therapeutic and the side-effects of NSAIDs. However, this theory has been refined since the discovery of a second, inducible COX isozyme, COX-2 [2, 3]. It has been proposed that COX-2 inhibition is relevant for the anti-inflammatory effects of NSAIDs, whereas inhibition of constitutive COX-1 is responsible for the gastric and renal side effects, as well as for the antithrombotic activity of these agents [4].

The suggestion that an inducible PGHS (for clarity, referred to as COX) existed, arose from studies in the field of inflammation and reproductive biology. As early as 1988, it was noticed that in spite of treatment with antibodies raised against COX purified from ram seminal vesicle (now known to be COX-1), fibroblasts were still able to increase their prostaglandin biosynthetic activity in response to interleukin-1 (IL-1) [5, 6]. Prostaglandin synthase activity was also shown to be strongly stimulated by bacterial endotoxin in human monocytes *in vitro* [2] as well as mouse macrophages *in vivo* [7]. Importantly, this increase was associated with the *de novo*, glucocorticoid-sensitive synthesis of new COX protein [8]. Studies in the rat ovary also provided evidence that two immunologically distinct forms of COX exist and that one of these two forms can be selectively regulated by hormones

Inducible Enzymes in the Inflammatory Response, edited by D.A. Willoughby and A. Tomlinson
© 1999 Birkhäuser Verlag Basel/Switzerland

[9–11]. The existence of an inducible COX was also observed in bovine tracheal cells [12].

The identification of a second COX gene occured independently from the afore-mentioned fields of research, in laboratories studying the regulation of gene expression in response to mitogens. Using a cDNA screen from chick embryo fibroblasts carrying the oncogene of Rous sarcoma virus, Xie et al. [13] identified a mRNA showing "significant sequence identity" with ovine COX mRNA. This message could be induced by serum or tumor promoters. The authors postulated that this message might be the product of a second COX gene, related to inducible COX. TIS10, a phorbol ester tumor promoter-inducible mRNA from Swiss 3T3 cells, was shown to also encode a prostaglandin synthase whose amino acid sequence was distinct from the previously cloned murine COX [3, 14]. Induction of this gene by mitogens could be prevented by dexamethasone [15, 16]. The first COX to be genetically cloned was the "classical" COX, from sheep seminal vesicle [17–19]. The cDNA for this COX-1 was later used to clone the murine [20, 21] and human genes [22–24]. COX-2 has been cloned from chicken [13, 25], mouse [14, 26], rat [21, 27–29] and human sources [30–32]. The three dimensional structures of COX-1 and COX-2 have also been determined [33–36].

Regulation of COX-2 expression in inflammatory cells *in vitro*

The strongest arguments supporting a role for COX-2 in inflammation are, first, that COX-2 expression can be induced by endotoxins, such as lipopolysaccharide (LPS), and pro-inflammatory cytokines in most cell types involved in inflammation. These include macrophages and monocytes, endothelial cells and synovial cells. Second, COX-2 expression is down-regulated in these cells by anti-inflammatory cytokines, such as IL-4, IL-10 and IL-13, as well as by glucocorticoids.

Up-regulation of COX-2 expression by LPS and pro-inflammatory cytokines

Induction of COX-2 by LPS has been demonstrated in several phagocytic cells, including monocytic cell lines [37–40], human monocytes [41–43], murine peritoneal and alveolar macrophages [44–48]. In these cell types, pro-inflammatory cytokines, such as IL-1β, interferon-gamma (IFN-γ) and tumor necrosis factor alpha (TNF-α) also induce COX-2 or prime the cells for full expression of COX-2 in response to other stimuli [38, 42, 49, 50]. In addition, IL-1β also induces COX-2 in other phagocytic cells, such as mesangial cells [51, 52], astrocytes and microglia [53, 54]. COX-2 expression in macrophages is also stimulated by other inflammatory mediators, such as platelet activating factor (PAF) [46] and prostaglandin E_2 (PGE$_2$) [47] and a close relationship exists between induction of COX-2 and inducible nitric oxide synthase (iNOS). This latter point is discussed in detail in the chapter by Buttery and Polak. Induction of COX-2 in macrophages involves reactive oxygen intermediates [55, 56] and protein tyrosine kinases [37, 57–60] and is accompanied by an increased prostaglandin biosynthesis. However, the statement that COX-2 is responsible for this induced-prostaglandin synthesis is challenged by observations in murine macrophages which suggest a more complex relationship between arachidonic acid metabolism and expression of COX isoforms [61, 62].

COX-2 can be induced in several other cell types involved in inflammatory processes, such as synoviocytes and endothelial cells. As early as 1987, using rabbit synoviocytes, Rothenberg [63] suggested that "cyclooxygenase is an inducible enzyme". Both COX-1 and COX-2 were later shown to be expressed in freshly explanted rheumatoid tissues but only COX-2 mRNA was markedly increased by treatment with IL-1β [64]. IL-1-induced prostaglandin synthesis was shown to be dependent on the *de novo* synthesis of COX-2 [65, 66]. Up-regulation of COX-2 expression by LPS, IL-1, IL-6 and TNF-α has also been reported in chondrocytes [66–68], osteoblasts [69] and synovial microvessel endothelial cells [70], suggesting a key role of this isoform in joint inflammation.

The vascular endothelium is involved in various stages of inflammation. An increased permeability due to retraction of endothelial cells is associated with the exudative process and the migration of phagocytic cells during acute inflammation. Exaggerated angiogenesis is a very important component of chronic inflammation. Induction of COX-2, but not COX-1, has been demonstrated in human umbilical vein endothelial cells [31, 71], human synovial microvessels [70] and rat brain vessels [72] in response to IL-1α and in bovine aortic endothelial cells in response to LPS, IL-1β, TNF-α, platelet derived growth factor (PDGF) and epidermal growth factor (EGF) [73]. Induction of COX-2 in endothelial cells, as seen in macrophages, also involves tyrosine kinases [73, 74]. Experiments on calf pulmonary artery endothelial cells suggested an increased expression of COX-1 rather than induction of COX-2 in response to transforming growth factor-β (TGF-β) and IL-1β [75].

COX-2 can also be induced by LPS and IL-1β in human gingival fibroblasts [76, 77], by IL-1β, PDGF, EGF, thrombin and following vessel injury in vascular smooth muscle cells [78, 79], by LPS, IL-1β, TNF-α, IFN-γ in pulmonary epithelial cells [80, 81], by IL-1β and TNF-α in airway smooth muscle cells [82, 83].

Down-regulation of COX-2 expression by anti-inflammatory cytokines and glucocorticoids

Recent evidence has indicated that IL-4, IL-10 and IL-13 have important regulatory effects on the immune and inflammatory systems. In some respects, these cytokines seem to have anti-inflammatory activities. This hypothesis fits well with the observation that they inhibit COX-2 expression under various experimental conditions. In human monocytes, IL-10 suppresses PGE_2-induced COX-2 whereas COX-1 is not affected. Attenuation of COX-2 expression is accompanied by decreased prostaglandin production [84]. In LPS-stimulated monocytes, IL-4, IL-10 and IL-13 inhibit prostanoid synthesis through suppression of inducible COX [85, 86]. Human recombinant IL-1 receptor antagonist, another natural anti-inflammatory peptide also inhibits COX-2 expression and PGE_2 production by monocyte cultures treated with LPS or IL-1β [87]. IL-4 and IL-13 are also able to inhibit bone resorption by suppressing COX-2 expression and related prostaglandin synthesis in osteoblasts stimulated by IL-1β [88]. In addition, IL-4 inhibits superantigen-induced PGE_2-dependent collagenase gene expression in human synoviocytes through inhibition of COX-2 and cytosolic phospholipase A_2 expression [89].

COX-2 induction can be inhibited by glucocorticoids in various cell systems and in response to various COX-2 inducers [41, 43, 44, 47, 52, 54, 64, 65, 67, 69, 70, 76, 77, 80, 82, 83]. The ability of glucocorticoids to down-regulate COX-2 expression may involve transcriptional [90] as well as post-transcriptional [91] mechanisms. Contrary to cytosolic phospholipase A_2 and iNOS, regulation of COX-2 expression by glucocorticoids at the transcriptional level is not mediated by lipocortin-1 [90, 92, 93]. A physiological role for endogenous glucocorticoids in the control of COX-2 expression has been demonstrated using adrenalectomized mice. Adrenalectomy causes elevation of COX-2 mRNA and protein, but not COX-1, in peritoneal macrophages. Dexamethasone replacement suppresses the elevation of COX-2 mRNA and protein, and the increase in COX enzyme activity observed in adrenalectomized animals [94].

COX-2 expression in inflamed tissue

Animal models of inflammation

Many animal models of inflammation have been developed for the discovery and evaluation of anti-inflammatory drugs. The most commonly used are carrageenan-induced paw edema in rats, air pouch model in mice or rats, adjuvant arthritis and experimentally induced pleurisy or peritonitis in rats and mice. COX-2 is not detectable in non-inflamed tissue, but is detectable after the injection of a pro-inflammatory stimulus in these animal models. COX-1 is unaffected by the inflammatory process, and similar levels of mRNA and protein are detectable in both normal and inflamed tissue in all of these models [95-101].

The time course of COX-2 appearance usually coincides with prostaglandin production and symptoms of inflammation. When carrageenan is injected into the paw of rats, a substantial induction of COX-2 is observed at 3 hours and this coincides with the appearance of thromboxane B_2 (TxB_2) and edema in the tissue [95]. This is also observed after the injection of carageenan into a subcutaneous air pouch, where the induction of COX-2 mRNA and protein coincides with the production of prostaglandins in the pouch tissue and cellular infiltrate. In the adjuvant arthritis model, high levels of COX-2 protein develop rapidly throughout the hindlimb joints, both contra- and ipsilateral to the site of Freund's adjuvant administration. This induction of COX-2 preceeds or parallels local production of PGE_2 and the symptoms of arthritis. Physiological doses of glucocorticoids suppress both arthritis and induction of COX-2 [100, 101].

In the air pouch model, dexamethasone inhibits both COX-2 expression and prostaglandin synthesis in the pouch exudate [96]. Induction of allergic inflammation by antigen challenge into the pouch is also accompanied by increased COX-2 levels. COX-1 levels remain unaffected. Dexamethasone inhibits the induction of COX-2 and suppresses the allergic inflammation [97]. In the mouse air pouch model of chronic granulomatous inflammation COX-2 mRNA and protein, as well as COX activity, increases over the first 24 hours post-injection and continues to rise for up to 14 days [98, 99]. COX-1 protein remains unchanged throughout.

In a rat carrageenan-induced pleurisy model of acute inflammation, COX-2 is the predominant enzyme in pleural exudate cells and COX activity peaks 2 to 6 hours after the injection of the pro-inflammatory stimulus [102]. In a model of chronic peritonitis, induced by the intraperitoneal injection of mineral oils, COX-2 expression is up-regulated in the macrophages isolated from peritoneal lavage samples [48]. COX-2 expression is also enhanced in a rat model of experimental glomerulonephritis induced by anti-glomerular membrane antibodies [58].

The observations that COX-2 but not COX-1 expression is induced in animal models of acute and chronic inflammation, and that prostaglandin biosynthesis and symptoms of inflammation parallel COX-2 induction, strongly support a predominant role for COX-2 in inflammatory processes [103]. However, observations in mice lacking the gene coding for COX-2 (knockout mice) do not concur with this theory. In fact, COX-2 knockout mice were shown to have a normal inflammatory response, at least in the mouse models of acute inflammation used [104–106]. This is discussed in more detail in another chapter of this book.

Human rheumatic diseases

Expression of COX in synovia from patients with rheumatoid arthritis (RA) and osteoarthritis (OA) was first studied by immunohistochemistry. In cells from RA synovia, extensive and intense intracellular COX immunostaining was observed, which correlated with the extent and intensity of mononuclear cell infiltration. Significantly less staining was observed in OA synovia [100]. Up-regulation of COX-2 in RA has been confirmed using immunoprecipitation and western blotting techniques [64, 107]. Studies using reverse transcription and the polymerase chain reaction suggested that both COX-1 and COX-2 may participate in prostaglandin biosynthesis in acute crystal and rheumatoid arthritis [108]. Similar conclusions were obtained by *in situ* hybridization studies in synovial tissues from patients with RA and OA [109].

Anti-inflammatory activity of COX-2 inhibition: pharmacological data

COX-2 appears to play a predominant role in inflammation, whereas COX-1 is involved in gastric cytoprotection, maintainance of kidney function and platelet aggregation. This statement is supported by pharmacological and early clinical data with selective inhibitors of COX-2. These data consistently show that selective COX-2 inhibitors have an anti-inflammatory activity similar to that of standard NSAIDs without the often seen side effects of these compounds [110].

A large number of *in vitro* assays have been developed to characterize the COX-1 and COX-2 inhibitory activities of NSAIDs. Depending on the test system, the experimental conditions vary greatly and different IC_{50} values (the concentration at which 50% of the activity is inhibited) and indices of selectivity are obtained. The index of selectivity is usually calculated in one of two ways. Either the IC_{50} of COX-1/IC_{50} of COX-2 is calculated, resulting in larger values for more selective COX-2 inhibitors (i.e. indomethacin 1, NS-398 100). Alternatively, the IC_{50} of COX-2/IC_{50} of COX-1 is calculated, resulting in the inverse (i.e. indomethacin 1, NS-398 0.01). Data using both methods have been published, and the latter method will be used in this publication (i.e. the smaller the value, the more COX-2 selective the substance).

In addition, comparisons of IC_{50} values and ratios of different compounds should only be performed when all were tested in the same system. Furthermore, the value of each test system should be taken into account when analyzing results. For example, selectivity results obtained in a human whole blood assay are probably more representative than results obtained using animal enzymes in an artificial milieu. Taking the above into account, the analysis of the results obtained with different compounds in different *in vitro* test systems indicate the following trends: (1) Standard NSAIDs are equally effective in inhibiting COX-1 and COX-2, or slightly more active on COX-1, (2) meloxicam, nimesulide and etodolac inhibit COX-2 preferentially but not exclusively and some COX-1 inhibitory activity is present, (3) developmental compounds or pharmacological tools such as flosulide (CGP 28238; Ciby-Geigy), DuP-697 (DuPont Co), NS-398 (Taisho Pharmac Co), SC58125 (Searle Monsanto) and L-745,337 (Merck Frosst) inhibit COX-2 selectively [111].

Meloxicam (Boehringer Ingelheim) is an example of a substance where drug development spanned the evolution of the COX-concept. This drug was first characterized *in vivo* in animal models, before the existence of COX-2 was known, and had potent anti-inflammatory and analgesic activity in all standard models of inflammation. Comparison of the ulcerogenic dose and the anti-inflammatory dose in a rat model of adjuvant arthritis indicated a superior therapeutic margin of meloxicam in comparison with standard NSAIDs [112–114]. A preferential inhibition of COX-2 relative to COX-1 was only later demonstrated and is likely to account for the improved pharmacological profile of the compound [115, 116]. Depending on the *in vitro* assay used to characterize meloxicam, COX-2/COX-1 ratios can vary between 0.003 and 0.8 [111, 117-120]. In a human whole blood assay, the COX-2/COX-1 ratio was found to be approximately 0.1 [111, 119].

Other preferential COX-2 inhibitors include nimesulide and etodolac. The *in vivo* pharmacology of nimesulide (Helsinn) was also known before COX-2 was characterized. Nimesulide was initially described as an atypical NSAID, exhibiting a potent anti-inflammatory activity in several models of inflammation, but with less gastrointestinal toxicity in the rat than standard NSAIDs [121-123]. A preferential inhibition of COX-2 was later demonstrated in several *in vitro* assays. Depending on the assay, COX-2/COX-1 ratios range from 0.001 to 0.8 [111, 117, 120, 124–129]. In a human whole blood assay, the COX-2/COX-1 ratio was 0.2 to 0.06 [119, 128]. Etodolac (Wyett Ayerst) was also described as an NSAID with a better safety profile than standard NSAIDs in a rat adjuvant arthritis model [130]. Etodolac was also later shown to preferentially inhibit COX-2 relative to COX-1 [120, 131, 132], the COX-2/COX-1 ratio was 0.1 in a human whole blood assay.

Other compounds with favourable safety profiles were later determined to be selective COX-2 inhibitors. Flosulide exhibited potent anti-inflammatory activity in rat adjuvant arthritis with improved gastrointestinal tolerability when compared with standard NSAIDs [133]. A selective inhibition of COX-2 was also later found in several *in vitro* assays [120, 125, 132, 134–136]. A COX-2/COX-1 ratio of 0.02 was obtained in the human whole blood assay [136]. DuP 697 was reported to be a potent inhibitor of paw swelling in adjuvant arthritis (ID_{50}: 0.2 mg/kg/day per os) without causing gastric ulcers at doses up to 400 mg/kg per os (single administration) [137]. Selective inhibition of COX-2 was later demonstrated in several test systems [125, 135, 138–140] and the COX-2/COX-1 ratio in the human whole blood assay was 0.05 [132]. NS-398 was also first characterized *in vivo* in the rat where it inhibited prostaglandin production in inflamed tissue more potently than in the gastric mucosa or the kidney [141]. In standard models of inflammation, it was almost as potent as indomethacin, and no gastric ulcerations were seen at single oral doses up to 1000 mg/kg [142]. COX-2 selectivity was later demonstrated in various *in vitro* assays [119, 125, 128, 129, 132, 135, 136, 138, 139, 140, 143, 144]. In the human whole blood assay, a COX-2 / COX-1 ratio between 0.006 to 0.09 was obtained [119, 128, 135].

SC58125 is a prototype of newly designed selective COX-2 inhibitors. Depending on the assays, the selectivity ratios are in the range 0.001 to 0.08 [95, 119, 120, 124, 129, 132, 135, 145]. In the human whole blood assay, COX-2/COX-1 ratios between 0.007 and 0.08 are obtained [119, 135]. SC58125 inhibits carrageenan-induced prostaglandin synthesis and paw edema and adjuvant arthritis with ED_{50}

values between 0.1 and 10 mg/kg p.o. in the rat. No inhibition of prostaglandin synthesis by the gastric mucosa and no signs of gastric toxicity are observed at single oral doses up to 10 and 600 mg/kg, respectively [95, 101, 146]. L-745, 337 exhibits COX-2/COX-1 selectivity ratios between 0.001 and 0.01, depending on the *in vitro* assay used [119, 120, 125, 132, 136, 140, 147]. A COX-2/COX-1 ratio of 0.004 was obtained in the human whole blood assay [119]. It inhibits carrageenan-induced edema [125, 148] and adjuvant arthritis [149, 150] with ED_{50} values between 0.2 and 3 mg/kg p.o. without causing gastric ulcerations at doses up to 40 mg/kg (single oral dose). No gastrointestinal bleeding is detected in a ^{51}Cr excretion assay in monkeys receiving doses of 10 mg/kg twice daily for 5 days [125, 148].

Taken as a whole, these pharmacological results strongly suggest that selective inhibition of COX-2 is responsible for the anti-inflammatory effects of NSAIDs without or with significantly less side effects on the gastrointestinal tract, supporting an exclusive role for COX-2 in inflammation. However, recent experimental results in the rat also suggest a protective function of COX-2 in ulcer healing [151, 152] and experimental colitis [153]. If this is confirmed in humans, it would exclude patients with ulceration and inflammatory bowel disease from selective COX-2 therapy, as well as from standard NSAID therapy. However, the relevance of this in patients still has to be determined [110].

Anti-inflammatory activity of COX-2 inhibition: clinical data

Innumerable clinical trials of NSAIDs in rheumatic diseases such as adult and juvenile RA, ankylosing spondylitis, gout and OA have confirmed the anti-inflammatory effects of these drugs. Clinical studies have failed to demonstrate significant differences in efficacy of currently available NSAIDs. According to the COX-2 and COX-1 selectivity hypothesis, it would be expected that COX-2 selective NSAIDs have equal anti-inflammatory potency as compared to standard NSAIDs, but have less effect on physiological prostaglandin production [4, 110, 154]. Thus, the anticipated benefit of a preferential or selective COX-2 inhibitor is an improved side effect profile mainly with respect to the gastrointestinal system, but also regarding the renal and the haemostatic system, with no loss of anti-inflammatory activity in comparison with traditional NSAIDs.

Clinical data are available for the preferential COX-2 inhibitors meloxicam (Boehringer Ingelheim), nimesulide (Helsinn) and etodolac (Wyeth Ayerst). Results from short-term clinical trials are now also available for the newly developed compounds flosulide (Ciba-Geigy), celecoxib (Searle Monsanto) and MK-966 (Merck Frosst). Many of these results are very recent, and full publications in peer-reviewed journals are not yet available. In order to review the latest information, results presented as posters and oral presentations at international congresses will also be discussed.

Meloxicam has now been launched in several countries for the treatment of OA and RA. The recommended doses in these indications are 7.5 and 15 mg once daily. Meloxicam is a preferential inhibitor of COX-2 *in vitro* and in animal experiments, and has COX-1 sparing activity in humans. In this latter study, the effects of 7.5 mg/day of meloxicam were compared with 25 mg of indomethacin, given three times daily, on platelet aggregation and serum TxB2 (which are exclusively COX-1 dependent) and urinary excretion of PGE_2 in human female volunteers [155]. Platelet aggregation and TxB_2 were almost completely blocked by indomethacin and urinary excretion of PGE_2 was reduced by ~50%. These parameters were not affected by meloxicam, thus meloxicam, at 7.5 mg/day, is COX-1 sparing in humans *in vivo*. Similar studies with 15 mg of meloxicam are ongoing [156].

Clinical results for meloxicam in inflammatory joint diseases are available in OA and RA. In a three week, double-blind, placebo-controlled trial in more than 400 patients, meloxicam (7.5 and 15 mg) proved to be effective in treating the signs and symptoms of RA [157]. In two further double-blind studies in patients with RA, meloxicam proved to be as effective as standard NSAIDs. One trial was a short-term, 3 week trial in 276 patients. Meloxicam (15 mg) was as effective as piroxicam (20 mg). The other trial was for 6 months in 379 patients and compared meloxicam (7.5 mg) with naproxen (750 mg). There was no difference in efficacy between the treatment groups [158].

The efficacy of meloxicam in OA has been assessed in several short- and long-term double-blind studies. In a placebo-controlled, three week trial in more than 400 patients, both 7.5 mg and 15 mg of meloxicam were not only superior to placebo, but also well tolerated, with an adverse event rate comparable to that of placebo [159]. In four comparative, multicentre trials in OA (2 short-term trials, 6 weeks and 2 long-term trials, 6 months) in a total of 1305 patients, 7.5 and 15 mg of meloxicam proved to be as effective as diclofenac (100 mg slow release) and piroxicam (20 mg) [160–163]. In all of these trials, there was a trend in favor of meloxicam for improved gastrointestinal tolerability as compared to diclofenac, piroxicam and naproxen. However, in most cases, this difference in each individual trial did not reach statistical significance.

The gastrointestinal tolerability of meloxicam was more thoroughly evaluated using a global analysis of the pooled safety data of double-blind studies in OA and RA, i.e. trials performed for drug registration in Europe [164]. Meloxicam, in doses of 7.5 mg and 15 mg (n = 893 and 3282 respectively), was compared with 20 mg piroxicam (n = 906), 100 mg slow release diclofenac (n = 324) and 750–1000 mg naproxen (n = 243). With respect to all gastrointestinal adverse events, severe (i.e. temporarily incapacitating) gastrointestinal adverse events and abdominal pain, both doses of meloxicam were significantly better tolerated than all comparators. Meloxicam was also safer than the comparators in reducing the frequency of upper gastrointestinal perforation, ulceration and bleeding. This reached statistical significance when compared to piroxicam and naproxen. This improved adverse event profile is consistent with results from endoscopy studies [165] as well as studies in patients with mild to moderate renal impairment [166].

These results have been confirmed in two large-scale double-blind studies over four weeks in patients with acute exacerbations of OA. The studies compared 7.5 mg meloxicam with 100 mg SR diclofenac (MELISSA Study [167]) and 20 mg piroxicam (SELECT Study [168], Boehringer Ingelheim, data on file) in 9323 and 8656 patients per study, respectively. They are the largest prospective double-blind head-to-head comparisons for NSAIDs and were designed to reflect clinical practice as closely as possible, i.e. including a significant percentage ($>40\%$) of patients older than 65. These two studies demonstrated comparable efficacy of 7.5 mg meloxicam, 100 mg SR diclofenac and 20 mg piroxicam.

In the MELISSA study, patients receiving meloxicam experienced significantly fewer gastrointestinal adverse events as compared to diclofenac (13% vs 19%; $p < 0.001$). The most commonly occurring gastrointestinal adverse events such as dyspepsia, nausea and vomiting, abdominal pain and diarrhoea were all significantly less frequent with meloxicam. Accordingly, the percentage of patients withdrawing from the study as a result of gastrointestinal adverse events was significantly lower with meloxicam (3.0%) than with diclofenac (6.1%, $p < 0.001$). Major and minor gastrointestinal ulcers, some of them associated with perforation or bleeding were seen in 5 and 7 patients in the meloxicam and diclofenac groups, respectively. Ulcer complications (2 perforated duodenal ulcers, 1 haemorrhagic duodenal ulcer and 1 bleeding gastric ulcer) were only seen in the diclofenac group. Hospitalization due to gastrointestinal adverse events occurred in 3 and 11 patients in the meloxicam and diclofenac groups, respectively. The mean duration of hospitalization in these patients was 2 and 11 days, respectively.

The results of the SELECT trial were similar. The incidence of gastrointestinal adverse events was significantly lower in the meloxicam than in the piroxicam group (10.3% vs 15.4%, $p < 0.001$). Individual gastrointestinal events such as dyspepsia, nausea/vomiting and abdominal pain occurred significantly less often with meloxicam than piroxicam. Major or minor bleeding, ulceration or perforation was observed in 16 patients in the piroxicam group as compared to 7 in the meloxicam group. Four complicated gastroduodenal ulcers occurred in the piroxicam group (1 bleeding gastric ulcer, 2 perforated gastric ulcers, 1 bleeding duodenal ulcer) while one such event (haematemesis with concurrent gastric and duodenal ulcer) occurred in the meloxicam group.

Clinically relevant increases of serum urea and serum creatinine were significantly more frequent in the diclofenac and piroxicam groups as compared to meloxicam, and relevant increases of liver transaminases were significantly more frequent with diclofenac as compared to meloxicam. There were also significantly fewer decreases in red blood cell count with meloxicam than with diclofenac and piroxicam, suggesting a higher frequency of occult gastrointestinal bleeding with the latter two drugs.

The outcome of these two large scale trials is consistent with the global safety analysis, indicating that meloxicam, even though it is not completely devoid of gastrointestinal side effects, has a better gastrointestinal tolerability than equally effective doses of standard, non-COX-2-selective NSAIDs. Taken as a whole, clinical trials with meloxicam suggest anti-inflammatory activity similar to other NSAIDs, but with improved gastrointestinal and renal tolerability. These data have now to be confirmed with post-marketing experience.

Nimesulide is commercially available in several countries for the management of OA. No clinical trials in RA have been published to date. The recommended doses are 100 and 200 mg given twice daily. Like meloxicam, nimesulide has consistently shown a preferential inhibition of COX-2 *in vitro*. However, results from human pharmacology studies are sometimes conflicting. Using an ex vivo whole blood assay (blood withdrawn from human volunteers who took the test compound for several days), either ~50% inhibition [119] or no inhibition [169] of serum TxB_2 (a marker of COX-1 activity *in vivo*) were reported in a dose of 100 mg bid. In an interaction study with furosemide, 200 mg nimesulide bid induced a transient decrease in indices of renal hemodynamics, markedly reduced urinary excretion of PGE_2 and attenuated the natriuretic, kaliuretic and diuretic effects of furosemide [170]. Since all these effects are characteristic for NSAIDs and are likely related to COX-1 inhibition, it seems that 200 mg nimesulide bid is not COX-1 sparing *in vivo* in humans. For 100 mg bid, the differences observed in the human whole blood assay may be explained by the steep plasma concentration curve and relatively high peak plasma levels of the compound. Concentrations which inhibit COX-1 to a significant extent may be temporarily reached at peak levels. Depending on the time points of blood sampling after drug administration, this effect may or may not be detected [111].

Nimesulide was investigated in various double-blind studies in patients with OA, the majority of them being short term trials in relatively small patient populations. No difference in efficacy or safety was detected with respect to 1000 mg naproxen, 600 mg etodolac, 20 mg piroxicam and 200 mg ketoprofen [171]. Three double-blind studies with larger patient numbers have also been performed. In a four week study in 392 patients with OA of the knee, 100 mg nimesulide bid appeared to be more effective than placebo, 200 mg nimesulide bid was as effective as 100 mg bid, but had a higher incidence of adverse events. It was concluded that 100 mg nimesulide bid is the optimal dose in the treatment of OA [172]. A double blind twelve week study in 199 patients with OA of the knee [173] compared 100 mg nimesulide bid with 300 mg etodolac bid. No significant difference in efficacy or safety between both drugs was detected. Nimesulide (100 mg bid) was also compared to naproxen (1000 mg) in 120 patients with OA of the hip in a study of four weeks duration [174]. Both drugs appeared to be equally effective and there was also no difference in safety.

Short-term clinical studies, most of them with small sample sizes, have investigated the anti-inflammatory effects of orally (mostly 100 mg bid) or rectally

(400 mg/day) administered nimesulide in a variety of painful inflammatory conditions such as bursitis/tendinitis, thrombophlebitis and upper airways inflammation. In most of these conditions, nimesulide was comparable in efficacy to standard NSAIDs [171].

In conclusion, available data from clinical trials in patients indicate that the efficacy and gastrointestinal tolerability of nimesulide is similar to that of other NSAIDs [171]. The gastrointestinal tolerability data are not in complete agreement with the results of two endoscopic trials in volunteers which demonstrated an improved gastrointestinal tolerability for 100 mg nimesulide bid as compared to 50 mg indomethacin tid [175, 176]. However, they have been confirmed by epidemilogical results in Italy, where the drug is widely prescribed, and which indicate that nimesulide does not have a better gastrointestinal tolerability than standard NSAIDs [154, 177].

Etodolac is registered for both OA and RA in several countries. Etodolac has shown a preferential inhibition of COX-2 in several *in vitro* assays. However, with the exception of a gastric biopsy trial (see below) [178], there are no human pharmacology studies investigating the effects of this compound on *in vivo* markers of COX-1 activity, such as serum TxB_2, platelet aggregation or urinary excretion of PGE_2. Etodolac is usually given twice daily. Clinical studies have established that the compound is as effective as standard NSAIDs in OA and RA [179]. A review of early clinical studies with etodolac in RA concluded that dosages of ≥ 400 mg are more effective than placebo and that daily doses of 600 mg of etodolac (300 mg bid) are comparable in efficacy to other NSAIDs [180]. In a double-blind study of 12 weeks duration including ~140 patients with RA per group [181], 600 mg etodolac was as effective but less ulcerogenic than 20 mg piroxicam, the 400 mg dose of etodolac was somewhat less effective.

In OA, 600 mg etodolac was compared in three larger (~200 and 300 patients in each trial) double-blind studies of 8 to 12 weeks duration to 150 mg diclofenac [181] and 20 mg piroxicam [182, 183]. No difference in efficacy was detected. Three double-blind, parallel group, placebo-controlled 4 week studies in patients with OA have compared 800 mg etodolac to 1000 mg naproxen and 1500 mg nabumetone [184]. In some efficacy endpoints, etodolac was significantly superior to nabumetone. No significant differences with respect to naproxen were observed. The gastrointestinal tolerability of etodolac was investigated in four endoscopic studies. In one study of seven days duration which included 72 healthy subjects (12 per group), the endoscopic scores in volunteers receiving 600 and 1000 mg of etodolac were not significantly different from placebo and significantly less than scores of those receiving 200 mg indomethacin, 1000 mg naproxen and 2400 mg ibuprofen [185]. In another endoscopic study, etodolac in doses between 400 and 1200 mg produced less gastroduodenal lesions than 3000 mg of aspirin [186]. No difference in gastroscopic scores was observed when 600 and 1200 mg of etodolac were compared with 150 mg diclofenac [187]. In a 4 week endoscopic study in patients with RA, 1000 mg naproxen, but not 600 mg etodolac appeared to suppress gastric and duodenal PGE_2 and PGI_2 levels in mucosal biopsy specimens, and naproxen also caused more mucosal lesions than etodolac. However, there was no correlation between PGE_2/PGI_2 levels and gastrointestinal damage in individual patients [178, 188]. In microbleeding studies, etodolac in doses ranging from 600 to 1200 mg for 1 to 4 weeks did not produce clinically significant bleeding in healthy volunteers or in arthritic patients, although in some studies faecal blood loss was significantly higher than in the placebo run-in period [179]. Feacal blood loss with etodolac was significantly lower in these studies than with anti-inflammatory doses of standard NSAIDs such as ibuprofen, indomethacin, piroxicam and naproxen.

In a study which was initially set up to investigate if etodolac or ibuprofen in anti-inflammatory doses could alter the progression of joint destruction in RA [189], no chondroprotective effect was observed. However, in the patients who completed one year of treatment, the crude rate of gastrointestinal ulcers and bleeding was more frequent with ibuprofen (2.2%) than with etodolac (0.4% and 0.5%, 300 and 1000 mg, respectively). Taken together these results suggest an improved gastrointestinal tolerability for etodolac over standard NSAIDs in equipotent doses. However, large scale head-to-head comparisons are needed to confirm these observations.

Flosulide has shown a high degree of selectivity for COX-2 in several *in vitro* assays. Unfortunately, its clinical development has been discontinued. In a double blind, crossover endoscopy study, flosulide (20 mg bid) was compared with naproxen (500 mg bid) in 19 patients with OA. Subjects were treated for 2 weeks with a two-week washout period. There was no difference in efficacy between the two compounds. However, using the Lanza scale as the primary endpoint to evaluate gastrointestinal damage, no damage was seen in 13 (68%) patients after flosulide and in 5 (37%) after naproxen (p < 0.001). Lanza scores were significantly lower with flosulide treatment as compared to naproxen, and flosulide was significantly better tolerated [190]. These results further support the idea that selective COX-2 inhibition is associated with a better gastrointestinal safety profile compared to traditional NSAIDs. Unfortunately, a placebo-controlled study investigating the renal effects of flosulide (25 mg/kg) in male normotensive volunteers revealed an increase in body weight, most likely related to a renal retention of water and electrolytes [191]. It remains to be determined whether the weight gain and influence on renal pharmacology observed with flosulide are related to the chemical structure of the compound or if they are related to the highly selective COX-2 inhibition of this compound.

Celecoxib (SC 58635) is a new COX-2 inhibitor in clinical development at Searle. It is the follow-up compound for SC 58125, whose development has been discontinued. Unfortunately, most of the preclinical pharmacology data available have been published with SC 58125, not with celecoxib. Using human recombinant enzymes, a COX-2/COX-1 selectivity ratio of 0.002 has been reported for celecoxib [192]. Results in a human whole blood assay have not yet been published. Nevertheless, in an *ex vivo* study in six healthy male volunteers [193, 194], celecoxib (600 mg bid) did not inhibit collagen-, ADP- or arachidonate-induced platelet aggregation and also had no effect on TxB_2 levels, two and four hours after the last dose on day six. This demonstrated a COX-1 sparing effect *in vivo*. In contrast, a 650 mg single dose of aspirin, which was used as a positive control significantly inhibited collagen- and arachidonate-induced platelet aggregation and TxB_2 levels. Celecoxib has been demonstrated to have analgesic effects in dental surgery studies [194]. According to its pharmacokinetic profile (half life 10 hours), it must be given twice daily [195].

A 4 week clinical study in 330 patients with active RA [194–196] compared 40, 200 and 400 mg bid of celecoxib with placebo. In the placebo group, 18% withdrew due to insufficient efficacy as compared to 5% and 8% of the patients given 400 or 800 mg celecoxib, respectively ($p < 0.05$ between both doses of celecoxib and placebo). Patients with the higher doses of celecoxib also showed markedly greater improvement than those in the placebo group regarding patient global assessment of efficacy, morning stiffness and number of painful/tender joints. Both doses were comparable to placebo regarding gastrointestinal adverse events. The recommended dose will probably be 400 mg bid for further trials in RA.

Celecoxib was compared to placebo in 293 patients with flares of OA of the knee at doses of 40, 100 and 200 mg bid. In this 2 week study, the two higher doses of celecoxib were of comparable efficacy and both were significantly more effective than placebo with regard to patient withdrawal due to lack of efficacy, patient global assessment and pain intensity. Both doses were comparable to placebo regarding gastrointestinal tolerability [197]. The recommended dose for further trials in OA will probably be 200 mg bid. An endoscopy trial was also performed to measure gatric ulceration and erosions after administration of placebo, 100 and 200 mg bid celecoxib and 500 mg bid naproxen. Erosions were seen in less than 20% of volunteers receiving placebo or celecoxib and in 75% of the volunteers receiving naproxen [196].

MK-966 is a novel COX-2 inhibitor from Merck Frosst which is currently under clinical investigation. It is a follow-up compound of L-745, 337, for which significantly more information on preclinical pharmacology is available. MK-966 has a long half-life allowing a daily application [198]. When using a human whole blood assay *ex vivo*, MK-966 was shown to be COX-1 sparing, i.e. it did not affect platelet TxB_2 production in doses up to 1 g/day. Analgesic effects were measured in patients after dental extraction at doses of 50 and 500 mg/day. The efficacy of MK-966 was not significantly different from ibuprofen 400 mg/day [199, 200]. In OA, doses of 25 and 125 mg of MK-966 were compared to placebo in 219 patients with knee OA in a study of six weeks duration [201]. Both doses of MK-966 showed significantly greater efficacy than placebo. There was no difference in efficacy between 25 and 125 mg of MK-966. Gastrointestinal tolerability was similar to placebo for both doses. Based on these studies, the recommended dose for subsequent trials in OA will probably be 25 mg/day. Gastrointestinal tolerability was further investigated in a one week endoscopy study. MK-966 (1 g/day) was compared with 2.4 g/day ibuprofen, 2.6 g/day aspirin and placebo. The percentages of volunteers with erosions were 15% with placebo, 15% with MK-966, 80% with ibuprofen and 95% with aspirin.

Conclusions

With the exception of some intriguing results in knockout mice, it seems that *in vitro* and *in vivo* pharmacological results, COX-2 expression in inflammatory tissues, as well as clinical data for preferential or selective inhibitors of COX-2 support the hypothesis that COX-2 inhibition is responsible for the anti-inflammatory activity of NSAIDs, whereas COX-1 inhibition is responsible for their gastrointestinal side effects. Post-marketing experience with preferential inhibitors of COX-2 and long-term clinical trials with novel COX-2 inhibitors are now necessary to completely validate this concept. An open question is whether highly selective inhibitors of COX-2 will have an incidence of gastrointestinal and/or renal side effects similar to placebo and if so, how much selectivity is required to provide an optimal therapeutic index.

References

1   Vane JR (1971) Inhibition of prostaglandin synthesis as a mechanism of action of aspirin-like drugs. *Nature* 231: 232–235
2   Fu JY, Masferrer JL, Seibert K, Raz A, Needleman P (1990) The induction and suppression of prostaglandin H2 synthase (cyclooxygenase) in human monocytes. *J Biol Chem* 265: 16737–16740
3   Fletcher BS, Kujubu DA, Perrin DM, Herschmann HR (1992) Structure of the mitogen-inducible TIS10 gene and demonstration that the TIS10-encoded protein is a functional prostaglandin G/H synthase. *J Biol Chem* 267: 4338–4344
4   Vane JR (1994) Towards a better aspirin. *Nature* 367: 215–216
5   Raz A, Wyche A, Siegel N, Needleman P (1988) Regulation of fibroblast cyclooxygenase synthesis by interleukin-1. *J Biol Chem* 263: 3022–3025
6   Raz A, Wyche A, Needleman P (1989) Temporal and pharmacological division of fibroblast cyclooxygenase expression into transcriptional and translational phases. *Proc Natl Acad Sci USA* 86: 1657–1661
7   Masferrer JL, Zweifel BS, Seibert K, Needleman P (1990) Selective regulation of cellular cyclooxygenase by dexamethasone and endotoxin in mice. *J Clin Invest* 86: 1375–1379

8   Masferrer JL, Seibert K, Zweifel BS, Needleman P (1992) Endogenous glucocorticoids regulate an inducible cyclooxygenase enzyme. *Proc Natl Acad Sci USA* 89: 3917–3921

9   Wong WYL, DeWitt DL, Smith WL, Richards JS (1989) Rapid induction of prostaglandin endoperoxide synthase induced by luteinizing hormone and cAMP is blocked by inhibitors of transcription and translation. *Mol Endocrinol* 3: 1714–1723

10   Wong WYL, Richards JS (1991) Evidence for two antigenically distinct molecular weight variants of prostaglandin H synthase in the rat ovary. *Mol Endocrinol* 5: 1861–1868

11   Sirois J, Richard JS (1992) Purification and characterization of a novel, distinct isoform of prostaglandin endoperoxide synthase induced by human chorionic gonadotrophin in granulosa cells of rat preovulatory follicles. *J Biol Chem* 267: 11586–11592

12   Rosen GD, Birkenmeier TM, Raz A, Holtzman MJ (1989) Identification of cyclooxygenase-related gene and its potential role in prostaglandin formation. *Biochem Biophys Res Commun* 164: 1358–1365

13   Xie W, Chipman JG, Robertson DL, Erikson RL, Simmons DL (1991) Expression of a mitogen-responsive gene encoding prostaglandin synthase is regulated by mRNA splicing. *Proc Natl Acad Sci USA* 88: 1692–1696

14   Kujubu DA, Fletcher BS, Varnum BC, Lim R, Herschman HR (1991) TIS10, a phorbol ester tumor promoter-inducible mRNA from Swiss 3T3 cells, encodes a novel prostaglandin synthase/cyclooxygenase homologue. *J Biol Chem* 266: 12866–12872

15   O'Banion MK, Sadowski HB, Winn V, Young DA (1991) A serum- and glucocorticoid-regulated 4-kilobase mRNA encodes a cyclooxygenase-related protein. *J Biol Chem* 266: 2361–2367

16   Kujubu DA, Herrschman HR (1992) Dexamethasone inhibits mitogen induction of the TIS10 prostaglandin synthase/cyclooxygenase gene. *J Biol Chem* 267: 7991–7994

17   DeWitt DL, Smith WL (1988) Primary structure of prostaglandin G/H synthase from sheep vesicular gland determined from the complementary DNA sequence. *Proc Natl Acad Sci USA* 85: 1412–1416

18   Merlie JP, Fagan D, Mudd J, Needleman P (1988) Isolation and characterization of the complementary DNA for sheep seminal vesicle prostaglandin endoperoxide synthase (cyclooxygenase). *J Biol Chem* 263: 3550–3553

19   Yokoyama C, Takai T, Tanabe T (1988) Primary structure of sheep prostaglandin endoperoxide synthase deduced from cDNA sequence. *FEBS Lett* 11: 371–386

20   DeWitt DL, El-Harth A, Kraemer SA, Andrews MJ, Yao EF, Armstrong RL, Smith WL (1990) The aspirin and heme binding sites of ovine and murine prostaglandin endoperoxide synthases. *J Biol Chem* 265: 5192–5198

21   Feng L, Sun W, Xia Y, Tang WW, Chanmugam P, Soyoola E, Wilson CB, Hwang D (1993) Cloning two isoforms of rat cyclooxygenase: differential regulation of their expression. *Arch Biochem Biophys* 307: 361–388

22   Yokoyama C, Tanabe T (1989) Cloning of human gene encoding prostaglandin endoperoxide synthase and primary structure of the enzyme. *Biochem Biophys Res Commun* 165: 888–894

23   Funk CD, Funk LB, Kennedy ME, Pong S, Fitzgerald GA (1991) Human platelet/ery-throleukemia cell prostaglandin G/H synthase: cDNA cloning, expression and gene chromosomal assignment. *FASEB J* 5: 2304–2312

24   Takahashi Y, Ueda N, Yoshimoto T, Yamamoto S, Yokoyama C, Miyata A, Tanabe T, Fuse I, Hattori A, Shibata A (1992) Immunoaffinity purification and cDNA cloning of human platelet prostaglandin endoperoxide synthase (cyclooxygenase). *Biochem Biophys Res Commun* 197: 494–500

25   Xie W, Merril JR, Bradshaw WS, Simmons DL (1993) Structural determination and promoter analysis of the chicken mitogen-inducible prostaglandin G/H synthase gene and genetic mapping of the murine homologue. *Arch Biochem Biophys* 300: 247–252

26   Fletcher BS, Kujubu DA, Perrin DM, Herschman HR (1992) Structure of the mitogen-inducible TIS10 gene and demonstration that the TIS10-encoded protein is a functional prostaglandin G/H synthase. *J Biol Chem* 267: 4338–4344

27   Kennedy BP, Chan CC, Culp SA, Cromlish WA (1993) Cloning and expression of rat prostaglandin endoperoxide synthase (cyclooxygenase)-2 cDNA. *Biochem Biophys Res Commun* 197: 494–500

28   Sirois J, Levy LO, Simmons DL, Richards JS (1993) Characterization and hormonal regulation of the promoter of the rat endoperoxide synthase 2 gene in granulosa cells. *J Biol Chem* 268: 7384–7385

29   Yamagata K, Andreasson KI, Kaufmann WE, Barnes CA, Worley PF (1993) Expression of a mitogen-inducible cyclooxygenase in brain neurones: regulation by synaptic activity and glucocorticoids. *Neurons* 11: 371–386

30   Hla T, Neilson K (1992) Human cyclooxygenase-2 cDNA. *Proc Natl Acad Sci USA* 89: 7384–7388

31   Jones DA, Carlton DP, McIntyre TM, Zimmerman GA, Prescott SM (1993) Molecular cloning of human prostaglandin endoperoxide synthase type II and demonstration of expression in response to cytokines. *J Biol Chem* 268: 9049–9054

32   Appleby SB, Ristimäki A, Neilson K, Narko K, Hla T (1994) Structure of the human cyclo-oxygenase-2 gene. *Biochem J* 302: 723–727

33   Picot D, Loll PJ, Garavito RM (1994) The X-ray crystal structure of the membrane protein prostaglandin H2 synthase-1. *Nature* 367: 243–249

34   Loll PJ, Garavito RM (1994) The isoforms of cyclooxygenase: structure and function. *Curr Opin Invest Drugs* 3: 1171–1180

35   Luong C, Miller A, Barnett J, Chow J, Ramesha C, Browner MF (1996) Flexibility of the NSAID binding site in the structure of human cyclooxygenase-2. *Nature Struct Biol* 3: 927–933

36   Browner MF (1996) X-ray crystal structure of human cyclooxygenase-2. In: N Bazan, J Botting, J Vane (eds): *New targets in inflammation: inhibitors of COX-2 and adhesion molecules.* Kluwer Academic Publishers, Dordrecht, 71–74

37   Akarasereenont P, Mitchell JA, Appleton I, Thiemermann C, Vane JR (1994) Involvement of tyrosine kinase in the induction of cyclo-oxygenase and nitric oxide synthase by endotoxin in cultured cells. *Br J Pharmacol* 113: 1522–1528

38 Arias-Negrete S, Keller K, Chadee K (1995) Proinflammatory cytokines regulate cyclooxygenase-2 mRNA expression in human macrophages. *Biochem Biophys Res Commun* 208: 582–589

39 Swierskosz TA, Mitchell JA, Warner TD, Botting RM, Vane JR (1995) Co-induction of nitric oxide synthase and cyclo-oxygenase: interactions between nitric oxide and prostanoids. *Brit J Pharmacol* 114: 1335–1342

40 Barrios-Rodiles M, Keller K, Belley A, Chadee K (1996) Nonsteroidal antiinflammatory drugs inhibit cyclooxygenase-2 enzyme activity but not mRNA expression in human macrophages. *Biochem Biophys Res Commun* 225: 896–900

41 Nüsing R, Ullrich V (1992) Regulation of cyclooxygenase and thromboxane synthase in human monocytes. *Eur J Biochem* 206: 131–136

42 Glaser KB, Lock YW (1995) Regulation of prostaglandin H synthase 2 expression in human monocytes by the marine natural product manoalide and scalaradial. *Biochem Pharmacol* 50: 913–922

43 Patrignani P, Santini G, Panara MR, Sciulli MG, Greco A, Rotondo MT, Giamberardino M, Maclouf J, Ciabattoni G, Patrono C (1996) Induction of prostaglandin endoperoxide synthase-2 in human monocytes associated with cyclo-oxygenase-dependent F2-isoprostane formation. *Brit J Pharmacol* 118: 1285–1293

44 Lee SH, Soyoola E, Chanmungam P, Hart S, Sun W, Zhong H, Liou S, Simmons D, Hwang D (1992) Selective expression of mitogen-inducible cyclooxygenase in macrophages stimulated with lipopolysaccharide. *J Biol Chem* 267: 25934–25938

45 O'Sullivan MG, Chilton FH, Huggins EM, McCall CE (1992) Lipopolysaccharide priming of alveolar macrophages for enhanced synthesis of prostanoids involves induction of a novel prostaglandin H synthase. *J Biol Chem* 267: 14547–14550

46 Thivierge M, Rola-Pleszczynski M (1995) Up-regulation of inducible cyclooxygenase gene expression by platelet-activating factor in activated rat alveolar macrophages. *J Immunol* 154: 6593–6599

47 Tordjman C, Coge F, Andre N, Rique H, Spedding M, Bonnet J (1995) Characterisation of cyclooxygenase 1 and 2 expression in mouse resident peritoneal macrophages *in vitro*; interactions of non steroidal anti-inflammatory drugs with COX2. *Biochim Biophys Acta* 1256: 249–256

48 Hinson RM, Williams JA, Shacter E (1996) Elevated interleukin 6 is induced by prostaglandin E2 in a murine model of inflammation: possible role of cyclooxygenase-2. *Proc Natl Acad Sci USA* 93: 4885–4890

49 Misko TP, Trotter JL, Cross AH (1995) Mediation of inflammation by encephalogenic cells: interferon induction of nitric oxide synthase and cyclooxygenase 2. *J Neuroimmunol* 61: 195–204

50 Quinn MR, Park E, Schuller-Levis G (1996) Taurine chloramine inhibits prostaglandin E2 production in activated raw 264.7 cells by post-transcriptional effects on inducible cyclooxygenase expression. *Immunol Lett* 50: 185–188

51 Martin M, Neumann D, Hoff T, Resch K, DeWitt DL, Goppelt-Struebe M (1994) Inter-

leukin-1-induced cyclooxygenase 2 expression is suppressed by cyclosporin A in rat mesangial cells. *Kidney Internat* 45: 150–158

52 Rzymkiewicz D, Leungang K, Baird N, Morrison AR (1994) Regulation of prostaglandin endoperoxide synthase gene expression in rat mesangial cells by interleukin-1β. *Am J Physiol* 266: F39–F45

53 Fiebich B, Biber K, Lieb K, van Calker D, Berger M, Bauer J, Gebicke-Haerter PJ (1996) Cyclooxygenase-2 expression in rat microglia is induced by adenosine A2a-receptors. *Glia* 18: 152–160

54 O'Banion MK, Miller JC, Chang JW, Kaplan MD, Coleman PD (1996) Interleukin-1β induces prostaglandin G/H synthase-2 (cyclooxygenase-2) in primary murine astrocyte cultures. *J Neurochem* 66: 2532–2540

55 Feng L, Xia Y, Garcia GE, Hwang D, Wilson CB (1995) Involvement of reactive oxygen intermediates in cyclooxygenase-2 expression induced by interleukin-1, tumor necrosis factor-α, and lipopolysaccharide. *J Clin Invest* 95: 1669–1675

56 Tetsuka T, Baier LD, Morrison AR (1996) Antioxidants inhibit interleukin-1-induced cyclooxygenase and nitric-oxide synthase expression in rat mesangial cells. *J Biol Chem* 271: 11689–11693

57 Corbett JA, Kwon G, Marino M, Rodi CP, Sullivan PM, Turk J, McDaniel ML (1996) Tyrosine kinase inhibitors prevent cytokine-induced expression of iNOS and COX-2 by human islets. *Am J Physiol* 270: C1581–C1587

58 Chanmugam P, Feng L, Liou S, Jang BC, Boudreau M, Yu G, Lee JH, Kwon HJ, Beppu T, Yoshida M et al. (1995) Radicicol, a protein tyrosine kinase inhibitor, suppresses the expression of mitogen-inducible cyclooxygenase in macrophages stimulated with lipopolysaccharide and in experimental glomerulonephritis. *J Biol Chem* 270: 5418–5426

59 Hwang D, Fischer NH, Jang BC, Tak H, Kim JK, Lee W (1996) Inhibition of the expression of inducible cyclooxygenase and proinflammatory cytokines by sesquiterpene lactones in macrophages correlates with the inhibition of MAP kinases. *Biochem Biophys Res Commun* 226: 810–818

60 Tetsuka T, Srivastava SK, Morrison AR (1996) Tyrosine kinase inhibitors, genistein and herbimycin A, do not block interleukin-1β-induced activation of NF-κB in rat mesangial cells. *Biochem Biophys Res Commun* 218: 808–812

61 Dolecki GJ, Rogers M, Lefkowith JB (1995) Discordance between macrophage arachidonate metabolic phenotype and the expression of cytosolic phospholipase A2 and cyclooxygenase. *Prostaglandins* 49: 397–414

62 Wilborn J, DeWitt DL, Peters-Golden M (1995) Expression and role of cyclooxygenase isoforms in alveolar and peritoneal macrophages. *Am J Physiol* 268: L294–L301

63 Rothenberg RJ (1987) Modulation of prostaglandin E$_2$ synthesis in rabbit synoviocytes. *Arthritis Rheumatism* 30: 266–273

64 Crofford LJ, Wilder RL, Ristimäki AP, Sano H, Remmers EF, Epps HR, Hla T (1994) Cyclooxygenase-1 and -2 expression in rheumatoid synovial tissues: effects of interleukin-1β, phorbol ester and corticosteroids. *J Clin Invest* 93: 1095–1101

65 Angel J, Berenbaum F, Le Denmat C, Nevalainen T, Masliah J, Fournier C (1994) Inter-leukin-1-induced prostaglandin E_2 biosynthesis in human synovial cells involves the activation of cytosolic phospholipase A_2 and cyclooxygenase-2. *Eur J Biochem* 226: 125–131

66 Knott I, Dieu M, Burton M, Houbion A, Remacle J, Raes M (1994) Induction of cyclooxygenase by interleukin 1: comparative study between human synovial cells and chondrocytes. *J Rheumatol* 21: 462–466

67 Geng Y, Blanco FJ, Cornelisson M, Lotz M (1995) Regulation of cyclooxygenase-2 expression in normal human articular chondrocytes. *J Immunol* 155: 796–801

68 Berenbaum F, Jacques C, Thomas G, Corvol MT, Béréziat G, Masliah J (1996) Synergistic effect of interleukin-1β and tumor necrosis factor α on PGE_2 production by articular chondrocytes does not involve PLA_2 stimulation. *Exp Cell Res* 222: 379–384

69 de Brum-Fernandes AJ, Laporte S, Heroux M, Lora M, Patry C, Ménard HA, Dumais R, Leduc R (1994) Expression of prostaglandin endoperoxide synthase-1 and prostaglandin endoperoxide synthase-2 in human osteoblasts. *Biochem Biophys Res Commun* 198: 955–960

70 Szczepanski A, Moatter T, Carley WW, Gerritsen ME (1994) Induction of cyclooxygenase II in human synovial microvessel endothelial cells by interleukin-1. *Arthritis Rheumatism* 37: 495–503

71 Habbib A, Créminon C, Frobert Y, Grassi J, Pradelles P, Maclouf J (1993) Demonstration of an inducible cyclooxygenase in human endothelial cells using antibodies raised against the carboxyl-terminal region of the cyclooxygenase-2. *J Biol Chem* 268: 23448–23454

72 Cao C, Matsumura K, Yamagata K, Watanabe Y (1996) Endothelial cells of the rat brain vasculature express cyclooxygenase-2 mRNA in response to systemic interleukin-1: a possible site of prostaglandin synthesis responsible for fever. *Brain Res* 733: 263–272

73 Akarasereenont P, Bakhle YS, Thiemermann C, Vane JR (1995) Cytokine-mediated induction of cyclo-oxygenase-2 by activation of tyrosine kinase in bovine endothelial cells stimulated by bacterial lipopolysaccharide. *Brit J Pharmacol* 115: 401–408

74 Blanco A, Habib A, Levy-Toledano S, Maclouf J (1995) Involvement of tyrosine kinases in the induction of cyclo-oxygenase-2 in human endothelial cells. *Biochem J* 312: 419–423

75 Jackson BA, Goldstein RH, Roy R, Cozzani M, Taylor L, Polgar P (1993) Effects of Transforming Growth Factor β and Interleukin-1β on expression of cyclooxygenase 1 and 2 and phospholipase A_2 mRNA in lung fibroblasts and endothelial cells in culture. *Biochem Biophys Res Commun* 197: 1465–1474

76 Yucel-Lindberg T, Ahola H, Nilsson S, Carlstedt-Duke J, Modéer T (1995) Interleukin-1β induces expression of cyclooxygenase-2 mRNA in human gingival fibroblasts. *Inflammation* 19: 549–558

77 Noguchi K, Shitashige M, Yanai M, Morita I, Nishihara T, Murota S, Ishikawa I (1996)

Prostaglandin production via induction of cyclooxygenase-2 by human gingival fibrob-lasts stimulated with lipopolysaccharides. *Inflammation* 20: 555–568

78 Pritchard KA, O'Banion MK, Miano JM, Vlasik N, Bhatia UG, Young DA, Stemerman MB (1994) Induction of cyclooxygenase-2 in rat vascular smooth muscle cells *in vitro* and *in vivo*. *J Biol Chem* 269: 8504–8509

79 Rimarachin JA, Jacobson JA, Szabo P, Maclouf J, Créminon C, Weksler BB (1994) Reg-ulation of cyclooxygenase-2 expression in aortic smooth muscle cells. *Arterioscler Thromb* 14: 1021–1031

80 Mitchell JA, Belvisi MG, Akarasereenont P, Robbins RA, Kwon OJ, Croxtall J, Barnes PJ, Vane JR (1994) Induction of cyclo-oxygenase-2 by cytokines in human pulmonary epithelial cells: regulation by dexamethasone. *Br J Pharmacol* 113: 1008–1014

81 Akarasereenont P, Thiemermann C (1996) The induction of cyclo-oxygenase-2 in human pulmonary epithelial cell culture (A549) activated by IL-1β is inhibited by tyro-sine kinase inhibitors. *Biochem Biophys Res Commun* 220: 181–185

82 Vadas P, Stefanski E, Wloch M, Grouix B, Van den Bosch H, Kennedy B (1996) Secre-tory non-pancreatic phospholipase A2 and cyclooxygenase-2 expression by tracheo-bronchial smooth muscle cells. *Eur J Biochem* 235: 557–563

83 Belvisi M, Saunders MA, Haddad EB, Hirst S, Yacoub MH, Barnes PJ, Mitchell JA (1997) Induction of cyclo-oxygenase-2 by cytokines in human cultured airway smooth muscle cells: novel inflammatory role of this cell type. *Br J Pharmacol* 120: 910–916

84 Mertz PM, DeWitt DL, Stetler-Stevenson WG, Wahl LM (1994) Interleukin 10 sup-pression of monocyte prostaglandin H synthase-2. *J Biol Chem* 269: 21322–21329

85 Niro H, Otsuka T, Tanabe T, Hara S, Kuga S, Nemoto Y, Tanaka Y, Nakashima H, Kita-jima S, Abe M et al (1995) Inhibition by interleukin-10 of inducible cyclooxygenase expression in lipopolysaccharide-stimulated monocytes: its underlying mechanism in comparison with interleukin-4. *Blood* 85: 3736–3745

86 Endo T, Ogushi F, Sone S (1996) LPS-dependent cyclooxygenase-2 induction in human monocytes is down-regulated by IL-13, but not IFN-γ. *J Immunol* 156: 2240–2246

87 Porreca E, Reale M, Di Febbo C, Di Gioacchino M, Barbacane RC, Castellani ML, Bac-cante G, Conti P, Cuccurullo F (1996) Down-regulation of cyclooxygenase-2 (COX-2) by interleukin-1 receptor antagonist in human monocytes. *Immunology* 89: 424–429

88 Onoe Y, Miyaura C, Kaminakayashiki T, Nagai Y, Noguchi K, Qing-Rong C, Seo H, Ohta H, Nozawa S, Kudo I et al. (1996) IL-13 and IL-4 inhibit bone resorption by sup-pressing cyclooxygenase-2-dependent prostaglandin synthesis in osteoblasts. *J Immunol* 156: 758–764

89 Mehindate K, Al-Daccak R, Aoudjit F, Damdoumi F, Fortier M, Borgeat P, Mourad W (1996) Interleukin-4, transforming growth factor β1, and dexamethasone inhibit super-antigen-induced prostaglandin E_2-dependent collagenase gene expression through their action on cyclooxygenase-2 and cytosolic phospholipase A_2. *Lab Invest* 75: 529–538

90 Newman SP, Flower RJ, Croxtall JD (1994) Dexamethasone suppression of IL-1β-induced cyclooxygenase 2 expression is not mediated by lipocortin-1 in A549 cells. *Biochem Biophys Res Commun* 202: 931–939

91 Ristimäki A, Narko K, Hla T (1996) Down-regulation of cytokine-induced cyclo-oxygenase-2 transcript isoforms by dexamethasone: evidence for post-transcriptional regulation. *Biochem J* 318: 325–331

92 Wu CC, Croxtall JD, Peretti M, Bryant CE, Thiemermann C, Flower RJ, Vane JR (1995) Lipocortin 1 mediates the inhibition by dexamethasone of the induction by endotoxin of nitric oxide synthase in the rat. *Proc Natl Acad Sci USA* 92: 3473–3477

93 Croxtall JD, Newman SP, Choudhury Q, Flower RJ (1996) The concerted regulation of cPLA$_2$, COX2, and lipocortin 1 expression by IL-1β in A549 cells. *Biochem Biophys Res Commun* 220: 491–495

94 Masferrer JL, Reddy ST, Zweifel BS, Seibert K, Needleman P, Gilbert RS, Herschman HR (1994) *In vivo* glucocorticoids regulate cyclooxygenase-2 but not cyclooxygenase-1 in peritoneal macrophages. *J Pharmacol Exp Ther* 270: 1340–1344

95 Seibert K, Zhang Y, Leahy K, Hauser S; Masferrer J, Perkins W, Lee L, Isakson P (1994) Pharmacological and biochemical demonstration of the role of cyclooxygenase 2 in inflammation and pain. *Proc Natl Acad Sci USA* 91: 12013–12017

96 Masferrer JL, Zweifel BS, Manning PT, Hauser SD, Leahy KM, Smith WG, Isakson PC, Seibert K (1994) Selective inhibition of inducible cyclooxygenase 2 *in vivo* is antiinflammatory and nonulcerogenic. *Proc Natl Acad Sci USA* 91: 3228–3232

97 Niki H, Tominaga Y, Watanabe-Kobayashi M, Mue S, Ohuchi K (1997) Possible participation of cyclooxygenase-2 in the recurrence of allergic inflammation in rats. *Eur J Pharmacol* 320: 193–200

98 Vane JR, Mitchell JA, Appleton I, Tomlinson A, Bishop-Bailey D, Croxtall J, Willoughby DA (1994) Inducible isoforms of cyclooxygenase and nitric-oxide synthase in inflammation. *Proc Natl Acad Sci USA* 91: 2046–2050

99 Appleton I, Tomlinson A, Mitchell JA, Willoughby DA (1995) Distribution of cyclooxygenase isoforms in murine chronic granulomatous inflammation. Implications for future anti-inflammatory therapy. *J Pathol* 176: 413–420

100 Sano H, Hla T, Maier JAM, Crofford LJ, Case JP, Naciag T, Wilder RL (1992) *In vivo* cyclooxygenase expression in synovial tissues of patients with rheumatoid arthritis and osteoarthritis and rats with adjuvant and streptococcal cell wall arthritis. *J Clin Invest* 89: 97–108

101 Anderson GD, Hauser SD, McGarity KL, Bremer ME, Isakson PC, Gregory SA (1996) Selective inhibition of cyclooxygenase (COX)-2 reverses inflammation and expression of COX-2 and interleukin 6 in rat adjuvant arthritis. *J Clin Invest* 97: 2672–2679

102 Tomlinson A, Appleton I, Moore AR, Gilroy DW, Willis D, Mitchell JA, Willoughby DA (1994) Cyclo-oxygenase and nitric oxide synthase isoforms in rat carrageenin-induced pleurisy. *Br J Pharmacol* 113: 693–698

103 Pairet M, Engelhardt G (1996) Distinct isoforms (COX-1 and COX-2) of cyclooxygenase: possible physiological and therapeutic implications. *Fund Clin Pharmacol* 10: 1–15

104 Dinchuk JE, Car BD, Focht RJ, Johnston JJ, Jaffee BD, Covington MB, Contel NR, Eng

VM, Collins RJ, Czerniak PM et al. (1995) Renal abnormalities and an altered inflammatory response in mice lacking cyclooxygenase II. *Nature* 378: 406–409

105 Morham SG, Langenbach R, Loftin CD, Tiano HF, Vouloumanos N, Jennette JC, Mahler JF, Kluckman KD, Ledford A, Lee CA et al (1995) Prostaglandin synthase gene disruption causes severe renal pathology in the mouse. *Cell* 83: 473–482

106 DeWitt D, Smith WL (1995) Yes, but do they still get headaches. *Cell* 83: 345–348

107 Crofford LJ. (1996) Expression and regulation of cyclooxygenase-2 in synovial tissues of arthritic patients. In: N Bazan, J Botting, J Vane (eds): *New targets in inflammation: inhibitors of COX-2 or adhesion molecules*. Kluwer Academic Publishers, Dordrecht, 83–91

108 Carreira PE, Iniquez MA, Pablos JL, Cabré F, Gomez-Reino JJ (1996) Detection of COX-1 and COX-2 isoforms in synovial fluid from patients with inflammatory joint diseases. *Arthritis Rheumatism* 39 (Suppl 9): Abstract 327

109 Franz JK, Hummel KM, Aicher WK, Petrow PK, Gay RE, Gay S (1996) In-situ detection of cyclooxygenase (COX) 1 and 2 mRNA in rheumatoid arthritis (RA) and osteoarthritis (OA) synovium. *Arthritis Rheumatism* 39 (Suppl 9): Abstract 1021

110 van Ryn J, Pairet M (1997) Selective cyclooxygenase-2 inhibitors: pharmacology, clinical effects and therapeutic potential. *Exp Opin Invest Drugs* 6: 609–614

111 Pairet M, van Ryn J, Mauz A, Schierok H, Diederen W, Türck D, Engelhardt G (1998) Differential inhibition of COX-1 and COX-2 by NSAIDs: a summary of results obtained using various test systems. In: J Vane, J Botting (eds): *Selective cyclooxygenase-2 inhibitors: pharmacology, clinical effects and therapeutic potential*. Kluwer Academic Publishers, Dordrecht, 27–46

112 Engelhardt G, Homma D, Schlegel K, Utzmann R, Schnitzler C (1995) Anti-inflammatory, analgesic, antipyretic and related properties of meloxicam, a new non-stroidal anti-inflammatory agent with favourable gastrointestinal tolerance. *Inflamm Res* 44: 423–433

113 Engelhardt G, Homma D, Schnitzler C (1995) Meloxicam: a potent inhibitor of adjuvant arthritis in the lewis rat. *Inflamm Res* 44: 548–555

114 Engelhardt G, Bögel R, Schnitzler C, Utzmann R (1996) Meloxicam: influence on arachidonic acid metabolism. Part 2. *In vivo* findings. *Biochem Pharmacol* 51: 29–38

115 Pairet M, Churchill L, Trummlitz G, Engelhardt G (1996) Differential inhibition of cyclooxygenase-1 (COX-1) and -2 (COX-2) by NSAIDs: consequences on anti-inflammatory activity versus gastric and renal safety. *Inflammopharmacology* 4: 61–70

116 Pairet M, Churchill L, Engelhardt G. (1996) Differential inhibition of cyclooxygenases 1 and 2 by NSAIDs. In: N Bazan, J Botting, J Vane (eds): *New targets in inflammation: inhibitors of COX-2 and adhesion molecules*. Kluwer Academic Publishers, Dordrecht, 23–38

117 Churchill L, Graham A, Shih CK, Pauletti D, Farina PR, Grob PM (1996) Selective inhibition of human cyclo-oxygenase-2 by meloxicam. *Inflammopharmacology* 4: 125–135

118 Engelhardt G, Bögel R, Schnitzler C, Utzmann R (1996). Meloxicam: influence on arachidonic acid metabolism. Part 1. *In vitro* findings. *Biochem Pharmacol* 51: 21–28

119 Patrignani P, Panara MR, Santini G, Sciulli MG, Padovano R, Cipollone F, Patrono C (1996) Differential inhibition of cyclooxygenase activity and prostaglandin endoperoxide synthase isozymes *in vitro* and *ex vivo* in man. *Prostaglandins Leukotrienes and Essential Fatty Acids* 55 (Suppl 1): Abstract P115

120 Riendeau D, Percival MD, Boyce S, Brideau C, Charleson S, Cromlish W, Ethier D, Evans J, Falgueyret JP, Ford-Hutchinson AW et al (1997) Biochemical and pharmacological profile of a tetrasubstituted furanone as a highly selective COX-2 inhibitor. *Br J Pharmacol* 121: 105–117

121 Swingle KF, Moore GGI (1984) Preclinical pharmacological studies with nimesulide. *Drugs Exp Clin Res* 10: 587–597

122 Carr DP, Henn R, Green JR, Böttcher I (1986) Comparison of the systemic inhibition of thromboxane synthesis, anti-inflammatory activity and gastro-intestinal toxicity of non-steroidal anti-inflammatory drugs in the rat. *Agents Actions* 19: 374–375

123 Velo GP (1991) The anti-inflammatory, analgesic and antipyretic activity of nimesulide in experimental methods. *Drug Invest* 3 (Suppl 2): 10–13

124 Huff R, Collins P, Kramer S, Seibert K, Koboldt C, Gregory S, Isakson P (1995) A structural feature of N-[2-8cyclohexyloxy)-4-nitrophenyl] methanesulfonamide (NS-398) that governs their selectivity and affinity for cyclooxygenase 2 (COX2). *Inflamm Res* 44 (Suppl 2): S145–S146

125 Prasit P, Black WC, Chan CC, Ford-Hutchinson AW, Gauthier JY, Gordon R, Guay D, Kargman S, Lau CK, Li CS et al (1995) L-745, 337: a selective cyclooxygenase-2 inhibitor. *Med Chem Res* 5: 364–374

126 Taniguchi Y, Ikesue A, Yokoyama K, Noda K, Debuchi H, Nakamura T, Toda A, Shimeno H (1995) Selective inhibition by nimesulide, a novel non-steroidal anti-inflammatory drug, with prostaglandin endoperoxide synthase-2 activity *in vitro*. *Pharmaceut Sci* 1: 173–175

127 Tavares IA, Bishai PM, Bennett A (1995) Activity of nimesulide on constitutive and inducible cyclooxygenases. *Arzneim Forsch/Drug Res* 45: 1093–1095

128 Young JM, Panah S, Satchawatcharaphong C, Cheung PS (1996) Human whole blood assays for inhibition of prostaglandin G/H synthases-1 and -2 using A23187 and lipopolysaccharide stimulation of thromboxane B_2 production. *Inflamm Res* 45: 246–253

129 Miralpex M, Camacho M, Lopez-Belmonte J, Canalias F, Beleta J, Palacios JM, Vila L (1997) Selective induction of cyclo-oxygenase-2 activity in the permanent human endothelial cell line HUV-EC-C: biochemical and pharmacological chracterization. *Br J Pharmacol* 121: 171–180

130 Martel RR, Klicius J (1982) Comparison in rats of the anti-inflammatory and gastric irritant effects of etodolac with several clinically effective anti-inflammatory drugs. *Agents Actions* 12: 295–297

131 Glaser K, Sung ML, O'Neill K, Belfast M, Hartman D, Carlson R, Kreft A, Kubrak D, Hsiao CL, Weichman B (1995) Etodolac selectively inhibits human prostaglandin G/H synthase 2 (PGHS-2) versus human PGHS-1. *Eur J Pharmacol* 281: 107–111

132 Wong E, DeLuca C, Boily C, Charleson S, Cromlish W, Denis D, Kargman S, Kennedy BP, Ouellet M, Skorey K et al (1997) Characterization of autocrine inducible prostaglandin H synthase-1 (PGHS-2) in human osteosarcoma cells. *Inflamm Res* 46: 51–59

133 Wiesenberg-Boettcher I, Schweizer A, Green JR, Mueller K, Maerki F, Pfeilschifter J (1989) The pharmacological profile of CGP 28238, a novel highly potent anti-inflammatory compound. *Drugs Exp Clin Res* 15: 501–509

134 Klein T, Nüsing R, Pfeilschifter J, Ulrich V (1994) Selective inhibition of cyclooxygenase 2. *Biochem Pharmacol* 48: 1605–1610

135 Brideau C, Kargman S, Liu S, Dallob AL, Ehrich EW, Rodger IW, Chan CC (1996) A human whole blood assay for clinical evaluation of biochemical efficacy of cylooxygenase inhibitors. *Inflamm Res* 45: 68–74

136 Cromlish L, Kennedy BP (1996) Selective inhibition of cyclooxygenase-1 and -2 using intact insect cell assays. *Biochem Pharmacol* 52: 1777–1785

137 Gans KR, Gabraight W, Roman RJ, Haber SB, Kerr JS, Schnidt WK, Smith C, Hewes WE, Ackerman NR (1990) Anti-inflammatory and safety profile of DuP 697, a novel orally effective prostaglandin synthesis inhibitor. *J Pharmacol Exp Ther* 254: 180–187

138 Copeland RA, Williams JM, Giannaras J, Nurnberg S, Covington M, Pinto D, Pick S, Trzaskos J (1994) Mechanism of selective inhibition of the inducible form of prostaglandin G/H synthase. *Proc Natl Acad Sci USA* 91: 11202–11206

139 Gierse JK, Hauser SD, Creely DP, Koboldt C, Rangwala SH, Isakson PC, Seibert K (1995) Expression and selective inhibition of the constitutive and inducible forms of human cyclo-oxygenase. *Biochem J* 305: 479–484

140 Kargman S, Wong E, Greig GM, Falgueyret JP, Cromlish W, Eithier D, Yergey JA, Riendeau D, Evans JF, Kennedy B et al (1996) Mechanism of selective inhibition of human prostaglandin G/H synthase-1 and -2 in intact cells. *Biochem Pharmacol* 52: 1113–1125

141 Futaki N, Arai I, Hamasaka Y, Takahashi S, Higuchi S, Otomo S (1993) Selective inhibition of NS-398 on prostanoid production in inflamed tissue in rat carrageenan-air-pouch inflammation. *J Pharm Pharmacol* 45: 753–755

142 Futaki N, Yoshikawa K, Hamasaka Y, Arai I, Higuchi S, Izuka H, Otomo S (1993) NS-398, a novel non-steroidal anti-inflammatory drug with potent analgesic and antipyretic effects, which causes minimal stomach lesions. *Gen Pharmac* 24: 105–110

143 Barnett J, Chow J, Ives D, Chiou M, Mackenzie R, Osen E, Nguyen B, Tsing S, Bach C, Freire J et al. (1994) Purification, characterization and selective inhibition of human prostaglandin G/H synthase 1 and 2 expressed in the baculovirus system. *Biochim Biophys Acta* 1209: 130–139

144 Futaki N, Takahashi S, Yokoyama M, Arai I, Higuchi S, Otomo S (1994) NS-398, a new anti-inflammatory agent, selectively inhibits prostaglandin G/H synthase/ cyclooxygenase (COX-2) activity *in vitro*. *Prostaglandins* 47: 55–59

145 Reitz DB, Li JL, Norton MB, Reinhard EJ, Collins JT, Anderson GD, Greggory SA, Koboldt CM, Perkins WE, Seibert K et al (1994) Selective cyclooxygenase inhibitors: novel 1, 2-diarylcyclopentenes are potent and orally active COX-2 inhibitors. *J Med Chem* 37: 3878–3881

146 Pairet M, Schmidberger C, Engelhardt G (1996) Effect of meloxicam and SC 58125 on adjuvant arthritis in the rat. *Rheumatology in Europe* 25 (Suppl 1): Abstract 445

147 Panara MR, Greco A, Santini G, Sciulli MG, Rotondo MT, Padovano R, di Giamberardino M, Cipollone F, Cuccurullo F, Patrono C et al (1995) Effects of the novel anti-inflammatory compounds, N-[2-(cyclohexyloxy)-4-nitrophenyl] methanesulphonamide (NS-398) and 5-methanesulphonamido-6-(2, 4-difluorothiophenyl)-1-indanone (L-745, 337) on the cyclo-oxygenase activity of human blood prostaglandin endoperoxide synthases. *Br J Pharmacol* 116: 2429–2434

148 Chan CC, Boyce S, Brideau C, Ford-Hutchinson AW, Gordon R, Guay D, Hill RG, Li CS, Mancini J, Penneton M et al (1995) Pharmacology of a selective cyclooxygenase-2 inhibitor, L-745, 337: a novel nonsteroidal anti-inflammatory agent with an ulcerogenic sparing effect in rat and nonhuman primate stomach. *J Pharmacol Exp Ther* 274: 1531–1537

149 Visco D, Widmer R, Shen F, Orevillo C, Christen A, Chung C, Leger S, Prasit P, Therien M, Wang Z et al (1996) The effects of a selective cyclooxygenase-2 (COX-2) inhibitor on adjuvant-induced arthritis in the rat. *Prostaglandins Leukotrienes and Essential Fatty Acids* 55 (Suppl 1): Abstract P202

150 Pairet M, Schmidberger C, van Ryn J, Püschner H (1997) Effect of indomethacin, meloxicam and L-745, 337 on adjuvant arthritis in the rat. ILAR Congress of Rheumatology, Singapore, 8–13 June, Abstract book p. 158

151 Kishimoto Y, Wada K, Nakamoto K, Kitano M, Kamisaki Y, Kawasaki H, Itoh T (1996) Development of the system to estimate expression level of COX-2 message in gastric injury induced by ischemia-reperfusion in rats. *Prostaglandins Leukotrienes and Essential Fatty Acids* 55 (Suppl 1): Abstract P203

152 Peskar BM, Stroff T, Halter F, Schmassmann A (1996) Effect of L-745, 337, a selective inhibitor of cyclooxygenase 2, on healing of chronic gastric ulcers in rats. *Prostaglandins Leukotrienes and Essential Fatty Acids* 55 (Suppl 1): Abstract 147

153 Reuter BK, Asfaha S, Buret A, Sharkey KA, Wallace JL (1996) Exacerbation of inflammation-associated colonic injury in rat through inhibition of cyclooxygnase-2. *J Clin Invest* 98: 2076–2085

154 Parnham M (1997) Selective COX-2 inhibitors. *DN&P* 10: 182–187

155 Stichtenoth DO, Wagner B, Frölich JC (1997) Effects of meloxicam and indomethacin on cyclooxygenase pathways in healthy volunteers. *J Invest Med* 45: 44–49

156 Frölich JC, Stichtenoth DO (1998) Renal side effects of NSAIDs: role of COX-1 and COX-2. In: J Vane, J Botting (eds): *Selective cyclooxygenase-2 inhibitors: pharmacology, clinical effects and therapeutic potential*. Kluwer Academic Publishers, Dordrecht, 87–98

157 Lemmel EM, Bolten W, Burgos-Vargas R, Platt P, Nissilä M, Sahlberg D, Björneboe O, Baumgartner H, Valat JP, Franchimont P et al (1997) Efficacy and safety of meloxicam in patients with rheumatoid arthritis. *J Rheumatol* 24: 282–90

158 Bolten W (1996) Clinical experience with meloxicam, a selective COX-2 inhibitor. In: N

Bazan, J Botting, J Vane (eds): *New targets in inflammation: inhibitors of COX-2 and adhesion molecules*. Kluwer Academic Publishers, Dordrecht 105–116

159 Lund B, Distel M, Bluhmki E (1994) A double-blind placebo controlled study of three different doses of meloxicam in patients with osteoarthritis of the knee. *Scand J Rheumatol Suppl* 98, Abstract 117

160 Goei The H, Lund B, Distel M, Bluhmki E (1995) A double-blind randomised trial comparing meloxicam with diclofenac in osteoarthritis. *Rheumatology in Europe* 24 (suppl 3): 325, Abstract E51

161 Hosie J, Distel M, Bluhmki E (1995) A six month, double-blind study comparing meloxicam 15 mg with piroxicam 20 mg in osteoarthritis. *Rheumatology in Europe* 24 (suppl 3): 325, Abstract E50

162 Hosie J, Distel M, Bluhmki E (1996) Meloxicam in osteoarthritis: a six-month, double-blind comparison with diclofenac sodium. *Br J Rheumatol* 35 (Suppl 1): 39–43

163 Lindén B, Distel M, Bluhmki E (1996) A double-blind study to compare the efficacy and safety of meloxicam 15 mg with piroxicam 20 mg in patients with osteoarthritis of the hip. *Br J Rheumatol* 35 (Suppl 1): 35–38

164 Distel M, Mueller C, Bluhmki E (1996) Global analysis of gastrointestinal safety of a new NSAID, meloxicam. *Inflammopharmacology* 4: 71–81

165 Patoia L, Santucci L, Furno M, Dionisi MS, Dell'Orso M, Romagnoli M, Satarinia A, Marini MG (1996) A 4-week double-blind parallel-group study to compare the gastrointestinal effects of meloxicam 7.5 mg, meloxicam 15 mg, piroxicam 20 mg and placebo by means of faecal blood loss, endoscopy and symptom evaluation in healthy volunteers. *Br J Rheumatol* 35 (suppl 1): 61–67

166 Bevis PJR, Bird HA, Lapham G (1996) An open study to assess the safety and tolerability of meloxicam 15 mg in subjects with rheumatic disease and mild renal impairment. *Br J Rheumatol* 35 (suppl 1): 56–60

167 Hawkey C, Kahan A, Steinbrück K, Alegre C, Baumelou E, Bégaud B, Dequeker J, Isomäki H, Littlejohn G, Mau J, Papazoglou S, and the international MELISSA Study Group (1998) Gastrointestinal tolerability of meloxicam compared to diclofenac in osteoarthritis patients. *Brit J Rheumatol; in press*

168 Dequeker J, Hawkey C, Kahan A, Steinbrück K, Alegre C, Baumelou E, Bégaud B, Isomäki H, Littlejohn G, Mau J, Papazoglou S, on behalf of the Select study group (1998) Improvement in gastrointestinal tolerability of the selective COX-2 inhibitor, meloxicam, compared with piroxicam: results of the safety and efficacy large scale evaluation of COX inhibiting therapies (SELECT) trial in osteoarthritis. *Brit J Rheumatol; in press*

169 Cullen L, Kelly L, Coyle D, Forde R, Fitzgerald D (1996) Selective suppression of COX-2 during chronic administration of nimesulide in man. *William Harvey Research Conference: Selective COX-2 Inhibitors; pharmacology, clinical effects and therapeutic potential*, Cannes, 20–21 March, Abstract P3

170 Steinhäuslin F, Munajo A, Buclin T, Macciochi A, Biollaz J (1993) Renal effects of nimesulide in furosemide-treated subjects. *Drugs* 46 (Suppl 1): 257–262

171 Davis R, Brogden RN (1994) Nimesulide. An update of its pharmacodynamic and phar-macokinetic properties, and therapeutic efficacy. *Drugs* 48: 431–454

172 Bourgeois P, Dreiser RL, Lequesne MG, Macciocchi A, Monti T (1994) Multi-centre double-blind study to define the most favourable dose of nimesulide in terms of effi-cacy/safety ratio in the treatment of osteoarthritis. *Eur J Rheumatol Inflamm* 14: 39–50

173 Lücker PW, Pawlowski C, Friedrich I, Faiella F, Magni E (1994) Double-blind, ran-domised, multi-centre clinical study evaluating the efficacy and tolerability of nimesulide in comparison with etodolac in patients suffering from osteoarthritis of the knee. *Eur J Rheumatol Inflamm* 14: 29–38

174 Quattrini M, Paladin S (1995) A double-blind study comparing nimesulide with naprox-en in the treatment of osteoarthrosis of the hip. *Clin Drug Invest* 10: 139–146

175 Cipollini F, Mecozzi V, Altilia F (1989) Endoscopic assessment of the effects of nime-sulide on the gastric mucosa: comparison with indomethacin. *Curr Ther Res* 45: 1042–1047

176 Marini U, Spotti D, Magni E, Monti T (1990) Double-blind endoscopic study compar-ing the effect of nimesulide and placebo on gastric mucosa of dyspeptic subjects. *Drug Invest* 2: 162–166

177 Garcia Rodriguez LA, Cattaruzzi C, Troncon MG, Agostinis L (1998) Risk of hospital-ization for upper gastrointestinal tract bleeding associated with ketorolac, other non-steroidal anti-inflammatory drugs, calcium antagonists, and other antihypertensive drugs. *Arch Intern Med* 158: 33–39

178 Taha AS, McLaughlin S; Holland PJ, Kelly RW, Sturrock RD, Russell RI (1990) Effect on gastric and duodenal mucosal prostaglandins of repeated intake of naproxen and etodolac in rheumatoid arthritis. *Ann Rheum Dis* 49: 354–358

179 Balfour JA, Buckley MMT (1991) Etodolac: A reappraisal of its pharmacology and ther-apeutic use in rheumatic diseases and pain states. *Drugs* 42: 274–299

180 Spencer-Green G (1997) Low dose etodolac in rheumatoid arthritis: a review of early studies. *J Rheumatol* 24 (suppl 47): 3–9

181 Lightfoot R (1997) Comparison of the efficacy and safety of etodolac and piroxicam in patients with rheumatoid arthritis. *J Rheumatol* 24 (suppl 47): 10–16

182 Bacon PA (1990) An overview of the efficacy of etodolac in arthritic disorders. *Eur J Rheumatol and Inflamm* 10: 22–34

183 Paulsen GA, Baigun S, de Figueiredo JG, de Freitas GG (1991) Efficacy and tolerability comparison of etodolac and piroxicam in the treatment of patients with osteoarthritis of the knee. *Curr Med Res Opin* 12: 40–41

184 Schnitzer TJ, Constantine G (1994): Etodolac in the treatment of osteoarthritis, recent studies. *J Rheumatol* 24 (suppl 47): 23–31

185 Lanza F, Rack MF, Lynn M, Wolf J, Sanda M (1987) An endoscopic comparison of the effects of etodolac, indomethacin, ibuprofen, naproxen, and placebo on the gastroin-testinal mucosa. *J Rheumatol* 14: 338–341

186   Lanza F, Panagides J, Salom IL (1986) Etodolac compared with aspirin: an endoscopic study of gastrointestinal tracts of normal volunteers. *J Rheumatol* 13: 299–303

187 Van Eeden A, Schotborgh RH, Tygat GNJ (1990) An endoscopic evaluation of the effects of etodolac and diclofenac on the gastric and duodenal mucosa. *Clin Ther* 12: 496–502

188 Russell RI (1990) Endoscopic evaluation of etodolac and naproxen and their relative effects on gastric and duodenal prostaglandins. *Rheumatol Int* 10: 17–21

189 Neustadt DH (1997) Double-blind evaluation of the longterm effects of etodolac versus ibuprofen in patients with rheumatoid arthritis. *J Rheumatol* 24 (suppl 47): 17–22

190 Bjarnason I, Macpherson A, Rotman H, Schupp J, Hayllar J (1997) A randomized, double-blind, crossover comparative endoscopy study on the gastroduodenal tolerability of a highly specific cyclooxygenase-2 inhibitor, flosulide, and naproxen. *Scand J Gastroenterol* 32:126–130

191 Brunel P, Hornych A, Guyene TT, Sioufi A, Turri M, Menard J (1995) Renal and endocrine effects of flosulide, after single and repeated administration to healthy volunteers. *Eur J Clin Pharmacol* 49: 193–201

192 Penning TD, Talley JJ, Berteshaw SR, Carter JS, Collins PW, Docter SD, Graneto MJ, Lee LF, Malecha JW, Miyashiro JM et al. (1997) Synthesis and biological evaluation of the 1, 5-diarylpyrazol class of cyclooxygenase-2 inhibitors: Identification of 4-[5-(4-Methylphenyl)-3-(trifluoromethyl)-1H-pyrazol-1-yl]benzenesulfonamide (SC-58735, celecoxib). *J Med Chem* 40: 1347–1365

193 Schwartz B, Hurley SM, Hubbard RC, Yu SS, Talwalker S, Isakson P, Geis GS (1997) A pilot study of the platelet effects of SC-58635, a novel COX-2 selective inhibitor. *ILAR Congress of Rheumatology*, Singapore, 8–13 June, Abstract Book p. 159

194 Lipsky PE (1997) COX-2 inhibition in patients with arthritis: Theory and practice. *APLAR Journal of Rheumatology* 1: 69–72

195 Lipsky PE (1997) Selective COX-2 inhibition in arthritis. *ILAR Congress of Rheumatology*, Singapore, 8–13 June

196 Isakson P, Zweifel B, Masferrer J, Koboldt C, Seibert K, Hubbard R, Geis S, Needleman P (1998) Specific COX-2 inhibitors: from bench to bedside. In: J. Vane, J Botting (eds): *Selective cyclooxygenase-2 inhibitors: pharmacology, clinical effects and therapeutic potential*. Kluwer Academic Publishers, Dordrecht, 127–133

197 Hubbard RC, Koepp R, Yu SS, Talwalker S, Geis GS, Weisenhutter CW, Makarowshi WS (1997) A pilot study of the efficacy of SC-58635, a novel COX-2 selective inhibitor, in treating osteoarthritis. *ILAR Congress of Rheumatology*, Singapore, 8–13 June, Abstract Book p. 46

198 Ford-Hutchinson AT (1998) New highly selective COX-2 inhibitors. In: J Vane, J Botting (eds): *Selective cyclooxygenase-2 inhibitors: pharmacology, clinical effects and therapeutic potential*. Kluwer Academic Publishers, Dordrecht, 117–125

199 Ehrich E, Dallob A, Van Hacken A, Depré M, DeLepeleire I, DeSchepper C, Gertz B (1996) Demonstration of selective COX-2 inhibition by MK-966 in humans. *Arthritis and Rheumatism* 39 (Suppl 9): Abstract 328

200 Ehrich E, Mehlisch D, Perkins S, Brown P, Wittreich J, Lipschutz K, Gertz B (1996) Efficacy of MK-966, a highly selective inhibitor of COX-2 in the treatment of postoperative dental pain. *Arthritis and Rheumatism* 39 (Suppl 9): Abstract 329

201 Ehrich E, Schnitzer T, Mcilwain H, Levy R, Wolfe F, Weisman W, Morrison B, Bolognese J (1997) MK-966, a highly selective COX-2 inhibitor, is effective in the treatment of osteoarthritis in a 6-week pilot study. *ILAR Congress of Rheumatology*, Singapore, 8–13 June, Abstract Book p. 47

Inducible nitric oxide synthase and inflammation

Adrian J. Hobbs and Salvador Moncada

Wolfson Institute for Biomedical Research, University College London, Rayne Institute, 5 University Street, London WC1E 6JJ, UK

Introduction

During both acute and chronic inflammatory processes profound physiological adaptations are triggered in an attempt to limit tissue damage and remove the pathogenic insult. Such mechanisms include local and systemic vasodilatation, pyresis and the activation of the immune system. These changes are regulated by a number of diverse mediators such as cytokines, bacterial products, (neuro)peptides and eicosanoids. Moreover, the considerable advances in the understanding of the role of nitric oxide (NO) as an inter- and intra- cellular messenger molecule have also established that NO plays an important role as a mediator of inflammation.

The inducible isoform of NO synthase (iNOS) is not constitutively present in cells, but expressed following exposure to inflammatory mediators akin to those described above; further, the release of NO from iNOS is comparatively greater than that from the constitutive isoforms. As such, it is believed that NO production by iNOS is the predominant source of this mediator during pathophysiological episodes. The regulation of iNOS expression is complex and the mechanisms by which NO modulates the inflammatory process are multi-faceted and are likely to be both protective and detrimental depending upon the local concentration of NO. This article will attempt to provide a brief overview of the role of iNOS in the inflammatory process, which will consist of two major aspects. First, the regulation of iNOS expression and activity and second, the pathophysiological actions of NO.

NO synthase: chromosomal location and molecular cloning of isoforms

Considerable information concerning the physiological and pathophysiological activity and regulation of NOS, and hence NO itself, has been derived from the molecular characterisation of the various isoforms of the synthetic enzyme. It is this aspect, therefore, which should be considered first to set the background to the important role NO plays in host defence and inflammation.

Although the L-arginine:NO pathway has been identified in many species including fish, birds, plants and bacteria, NOS has been best studied in mammals [1, 2]. Based upon several criteria including cellular location, regulation of activity and substrate/inhibitor profiles, the NOS enzyme can be divided into three distinct isoforms. First, a constitutive form, whose activity is regulated by Ca^{2+} and calmodulin and which is found in neural tissue, both centrally and peripherally (nNOS or NOS I). A second Ca^{2+}/calmodulin requiring constitutive enzyme is present in vascular endothelial cells (eNOS or NOS III). Thirdly, a Ca^{2+}/independent isoform (iNOS or NOS II) which can be isolated from a number of cell types following induction with specific cytokines. Each exists as a homodimer of approximately 260 kDa; only the dimeric forms exhibit catalytic activity. Distinct genes encode each isoform of NOS and consist of either 26 exons (iNOS and eNOS) or 29 exons (nNOS) [1, 2]. The isoforms share 50–60% homology and display many similarities to the cytochrome P450 family of enzymes; indeed, it is this analogy that has provided many clues to the structural composition/arrangement and catalytic process mediated by NOS.

Sequence analysis of the predicted primary structure of NOS isoforms has distinguished important functional motifs. All NOS proteins possess a bi-domain structure and dimerisation to homodimers is required for enzymatic activity. The C-terminal half of the NOS protein bears remarkable resemblance to only one other mammalian protein, cytochrome P450 reductase (CPR), and appears to possess the same co-factor binding sites; consequently, this is often referred to as the reductase domain. At the extreme C-terminus is a nicotinamide adenine dinucleotide phosphate (NADPH) binding region, which is conserved in all NOSs and aligns perfectly with CPR [3]. The NADPH binding site is followed, in turn, by FAD and FMN consensus sequences [4]. NOS, unlike CPR, is a self-sufficient enzyme in that the oxygenation of substrate, L-arginine, occurs at a haem-site in the N-terminal portion (oxygenase domain) of the protein. Stoichiometric amounts of haem are present in NOS and are required for catalytic activity [5, 6]. Resonance Raman spectroscopy has demonstrated the haem coordination to be pentavalent with a thiolate (fifth) axial ligand [7]. Haem coordination is thought to be provided by Cys-415 (in nNOS) based upon homology to cytochromes P450 and haem incorporation following site-directed mutagenesis [8]. Indeed, this Cys residue is conserved in all NOS isoforms across differing species and corresponds to Cys-200 in human iNOS and Cys-184 in human eNOS. As expected, close to the haem (catalytic) site is a substrate (L-arginine) binding site, providing further evidence that L-arginine oxidation occurs at the haem iron [7]. Separation of the two domains (via trypsin cleavage at a specific alanine residue at the C-terminal end of the oxygenase domain) [9] enabled L-arginine and pterin binding sites to be located to the oxygenase domain. Bridging the reductase and oxygenase domains is a calmodulin binding site which appears to act as a switch to regulate electron flow between the two regions [10]. Finally, eNOS possesses consensus sequences for myristoylation/palmitoylation at its N-terminus, which confers its particulate nature. Post-translational modification at these sites is necessary for its membrane association, which has been shown to occur preferentially to the caveolae [11].

NO is generated via a five-electron oxidation of a terminal guanidinium nitrogen on the amino acid L-arginine. This stereospecific reaction is both oxygen- and NADPH-dependent and yields the co-product L-citrulline in addition to NO. The mechanism of NOS-catalyzed oxidation of L-arginine to NO proceeds in at least two distinct steps. The initial reaction involves N-hydroxylation of the guanidinium nitrogen to form N-hydroxy-L-arginine, which is the only intermediate identified to date [12]. The reaction utilizes one equivalent of NADPH and O_2 to conduct a simple two electron oxidation of nitrogen, and mimics a classical P450-like hydroxylation, common in the biotransformation of many nitrogen-containing xenobiotics [13]. However, subsequent steps in the conversion of N-hydroxy-L-arginine to NO and L-citrulline remain unclear. Importantly, if NO is the product, NOS must facilitate an odd-electron oxidation, as indicated by the requirement for 0.5 equivalents of NADPH for this process [12]. Such a transformation is therefore difficult to reconcile with cytochrome P450 ($2e^-$) chemistry. Significantly, oxidation of comparative N-hydroxy-guanidines has shown that nitroxyl (HNO), and not NO, is the preferred nitrogen-oxide product [14, 15]. Indeed, this is not surprising since HNO would be the expected two electron oxidation product from N-hydroxy-guanidines. Therefore, it is conceivable that NOS generates HNO from L-arginine, in a four electron process (entirely consistent with cytochrome P450 enzymology), and that a subsequent $1e^-$ oxidation of this product yields NO (which likely represents the active biological entity). Recent studies have supported this potential mechanism of catalysis. First, HNO has been shown to possess biological activity indistinguishable from NO, which seems attributable to the rapid conversion of HNO to NO by a variety of physiologically relevant oxidants including superoxide dismutase (SOD), oxygen and haemoproteins. Second, and in an analogous fashion, SOD has been demonstrated to enhance directly the formation of free NO from L-arginine by NOS [16]. The mechanism responsible for this potentiation was shown to be independent of superoxide anion ($O_2{}^{.-}$) dismutation and NOS activation. Thus, SOD appears to accelerate the conversion of an intermediate in the L-arginine:NO pathway, such as HNO, to NO, by a $1e^-$ oxidation similar to that described above. Finally, in a study analysing the release of free NO from purified NOS enzyme [17], NO could not be detected from reaction mixtures containing appropriate co-factors and substrate L-arginine unless SOD was added to the reaction mixture; the activity of SOD was shown to be due to the conversion of HNO to NO, since N_2O could be detected in the headspace gas.

Constitutive NOS and the inflammatory process

The role of constitutive NOS in inflammation has been extensively studied, and appears to be important in the early phases of infection and inflammation. This topic is somewhat outside the scope of the present article, but will be summarised very briefly below.

Acute inflammation is a temporary, defensive reaction to a trauma or pathogenic insult. Initially, 'acute phase' mediators including histamine, 5-HT, bradykinin and substance P are released, accompanied by platelet aggregation, complement activation and eicosanoid formation, the latter of which helps sustain the inflammatory process. Finally, polymorphonuclear cells infiltrate the surrounding tissue in response to chemotactic signals. NO is thought to be involved in many aspects of this response, but it is likely that constitutive NOS, rather than iNOS, is the source of this NO. Furthermore, the low amounts of NO released by the constitutive isozymes may have both inflammatory and anti-inflammatory actions. First, acute 'phase mediators' release NO from the vascular endothelium which causes vasodilatation and increased vascular permeability, which undoubtedly represent pro-inflammatory events [18]. Neurogenically-induced inflammation following activation of afferent nociceptive fibres and the release of substance P has a NO-mediated component which is characterised by vasodilatation and oedema formation and which can be both endothelium-dependent and endothelium-independent [19]. Nevertheless, NO also inhibits platelet aggregation and synergises markedly with prostacyclin in this respect [20]. Further, NO appears to be able to inhibit the extravasation of inflammatory cells by inhibiting cell adhesion to the vascular endothelium, an effect possibly mediated by inhibiting the expression of adhesion molecules on endothelial cells [21]. Certainly, the latter two actions constitute an anti-inflammatory function and consequently, the role of NO derived from constitutive NOS isoforms in acute inflammatory responses remains in question.

Inducible NOS and the inflammatory process

In comparison to constitutive NOS isozymes, it is inducible NOS which is thought to mediate the vast majority of pathophysiological effects attributed to NO and it is this isoform which is consequently believed to be of fundamental importance to the process of inflammation, certainly in chronic episodes involving the recruitment of mononuclear cells. The contribution of NO derived from iNOS to the inflammatory process has two facets; first, the regulation of iNOS expression is crucial to the high-output production of NO for pathological purposes and second, the role NO plays in the inflammatory process itself, which encompasses both a direct action and an indirect effect on the expression of other inflammatory proteins.

Promoter sequences and control of NOS expression

Although much has been deduced about the molecular composition of NOS, rather less is known about its regulation of expression. This is of extreme importance in the case of iNOS which, unlike eNOS and nNOS, is not expressed constitutively but induced by the influence of inflammatory mediators and bacterial products. Since iNOS possesses tightly bound calmodulin, it is independent of free calcium concentrations and, therefore, once expressed continues to synthesise NO indefinitely. Hence, iNOS activity is regulated by protein expression rather than functional modulation. The 5'-flanking region of human iNOS possesses approximately 66% homology to its murine counterpart [22] and both contain conserved consensus sequences for nuclear factor kappa-B (NF-κB), γ-interferon responsive elements (γ-IRE) and a tumour necrosis factor responsive element (TNF-RE). An A-activator-binding site (AABS) also exists in the promoter region, which is thought to convey liver-specific gene expression [23]. The murine iNOS promoter (like the human equivalent) possesses an NF-κB binding site at position −76 to −85 which has been shown to bind proteins of the Rel/NF-κB family in response to lipopolysaccharide (LPS) [24]. Further upstream, (position −913− to −923) is a γ-IRE which has been shown to bind interferon-response factor (IRF)-1 upon stimulation of mouse peritoneal macrophages (RAW 264.7 cells) with interferon (IFN)-γ [25]. The latter element appears to function as an enhancer and is not able to initiate gene expression independently [26]. However, this promoter sequence arrangement can be reconciled with the inability of IFN-γ alone to induce iNOS expression and to its marked synergism with LPS. In accord with this observation, it is thought that the DNA binding protein NF-κB may impose the greatest degree of transcriptional control on iNOS expression, at least in murine systems; this is discussed in greater detail below. Despite these similarities, the transcriptional control of murine and human iNOS expression is markedly different. In the murine system, it appears that a proximal 1.6 kb 5'-flanking region contains the necessary promoter sequences to induce full gene expression. However, in the human, the corresponding sequence produces little or no gene expression, despite possessing transcription factor consensus sequences. Indeed, if a 16 kb fragment upstream from the coding sequence is cloned, linked to a luciferase reporter gene and transfected into human cell lines, it is still insufficient to promote full gene expression (as measured by luciferase activity) [27]. Thus, there is a distinct difference in the requirements for iNOS induction between mouse and human; this may well reflect evolutionary changes in the manner in which NO is utilised to fight pathogenic insults in man. The differences may also explain the reported difficulty in inducing iNOS expression in many human cell lines [28].

NF-κB and the regulation of iNOS expression

NF-κB is a higher eukaryotic transcription factor that plays a pivotal role in the expressional regulation of many genes that encode pro-inflammatory proteins [29, 30]. NF-κB is a heterodimeric protein composed most frequently of the DNA-binding proteins p50 and p65 (or Rel-A), and resides in the cytoplasm as a latent form stabilized by an inhibitory protein termed IκB [30–32]. As a consequence of cellular activation by a plethora of pathophysiological mediators, including cytokines and bacterial products, the IκB protein undergoes phosphorylation, which tags the IκB rendering it a high-affinity substrate for a ubiquitin conjugating enzyme system. In turn, ubiquitination identifies IκB as a substrate for the proteasome, a Ca^{2+}-independent and non-lysosomal multicatalytic proteinase complex. Thus, phosphorylation-regulated ubiquitination results in a rapid degradation of IκB leading to liberation of the NF-κB heterodimer [33, 34]. This separation reveals nuclear localization and DNA-binding sequences which permit the NF-κB to translocate into the nucleus and bind to appropriate regulatory elements in promoter regions on target genes, resulting in gene expression and subsequent protein synthesis [32]. This process is transient and is terminated through a delayed induction of IκB expression by NF-κB itself [35]. The majority of stimuli that activate NF-κB are associated with oxidative stress, as manifested by increased production of reactive oxygen species (ROS), including superoxide (O_2^-), hydrogen peroxide (H_2O_2) and hydroxy radical (HO·) [36]. Experimental evidence suggests that there is a convergence of distinct stimuli to generate a common ROS, possibly H_2O_2, that serves as an intracellular second messenger in mediating NF-κB activation [37]. It is unknown, however, how these two pathways are coupled.

Since NO has been known for many years to play an important role in host defence [38], it was not surprising when agents known to interfere with NF-κB activity were shown to modulate the induction of iNOS. Firstly, many antioxidants including pyrrolidinedithiocarbamate (PDTC) and diethyldithiocarbamate (DETC) were demonstrated to inhibit iNOS expression in cultured cells [39, 40], in addition to non-selective protease inhibitors such as Nα-p-tosyl-L-phenylalanine (TPCK) and Nα-p-tosyl-L-lysine-chloromethylketone (TLCK) and more selective calpain (α-N-acetyl-Leu-Leu-norleucinal) and proteasome inhibitors (Z-IE(O-t-Bu)A-Leucinal). Each is believed to prevent the degradation of IκB by the proteasome and thereby prevent liberation of the active NF-κB complex [41, 42]. Several other distinct classes of chemical agent have been demonstrated to prevent expression of iNOS via inhibition of the NF-κB transduction system. Glucocorticoids such as dexamethasone interfere with iNOS expression in many cell types by reducing the DNA binding capacity of NF-κB [43–46]. Glucocorticoids have also been shown to induce the expression of the IκB protein, thereby preventing the liberation of free NF-κB heterodimer [47]. Non-steroidal anti-inflammatory drugs, including aspirin and sodium salicylate, also appear to down-regulate iNOS expression by interfering with NF-κB activity [48, 49].

Recently, it has become apparent that NO itself possesses the ability to modulate NF-κB activity. This not only has important implications for the feedback regulation of iNOS expression, but since NF-κB is known to regulate transcriptionally the expression of numerous pro-inflammatory proteins (see above), NO may act as a widespread modulator of immune function and host defence. It has been recently demonstrated (A.J. Hobbs and L.J. Ignarro, unpublished observations) that NO can both enhance and inhibit NF-κB activity and iNOS expression in response to LPS in cultured mouse peritoneal macrophages (RAW 264.7 cells). At lower concentrations, the presence of NO markedly potentiates subsequent NO production (as determined by measurement of stable oxidation products NO_2^-, NO_3^-), whereas at higher concentrations of NO there is a pronounced inhibition. Utilisation of electrophoretic mobility shift assays and northern blot analysis to assess NF-κB activity and iNOS mRNA levels, respectively, confirms the biphasic activity profile of NO. Thus, NO can indirectly modulate iNOS expression (and hence its own synthesis) by interfering with the NF-κB transduction pathway. The mechanisms responsible for this biphasic effect are not clear. Certainly, if reactive oxygen species are responsible for activation of NF-κB, then it could be envisaged that NO may be able to react with (since it is a free radical) and inactivate such species to reduce the degree of activation. Furthermore, it has been demonstrated that NO is able to induce the expression of IκB and inhibit the DNA-binding capacity of NF-κB [50, 51]. However, the mechanism(s) underlying NO-induced enhancement of NF-κB activity are rather more obscure. One possibility is that NO may inhibit certain antioxidant enzymes and consequently increase the concentration of ROS in the local environment which in turn would elicit a greater degree of NF-κB activation. One putative target in this regard is catalase, which rapidly degrades H_2O_2 to O_2 and H_2O; a reduction in catalase activity by NO binding to its ferric catalytic site would raise H_2O_2 levels and possibly stimulate NF-κB activity. Although these hypotheses are somewhat speculative, the potential for NO to play a pivotal role in regulating immune function has been demonstrated and remains to be explored.

Substrate and co-factor availability as regulators of iNOS activity

One essential co-factor for the synthesis of NO by NOS is tetrahydrobiopterin (BH_4) and it is GTP cyclohydrolase I (GTPCHI), which converts GTP to dihydro-neopterin, which is the rate-limiting synthetic enzyme. In many cell types, it has been demonstrated that expression of active iNOS protein only occurs if GTPCHI is co-induced, presumably to provide the necessary BH_4 for NOS catalytic activity [52, 53]. 2,4-diamino-6-hydroxypyrimidine (DAHP), a selective inhibitor of GTPCHI, completely suppresses the elevated levels of BH_4 in response to cytokines in vascular smooth muscle cells. This suggests that it is *de novo* synthesis of BH_4 which supports the high-output production of NO by iNOS, rather than the pterin 'salvage' pathway which converts sepiapterin to BH_4 [54]. Such observations may have important implications for the treatment of conditions such as septic shock, in which inappropriate iNOS activity results in excessive NO production and profound hypotension; in this case, inhibitors of BH_4 synthesis (e.g. DAHP, methotrexate and N-acetylserotonin) may be able to reduce NO production by limiting the supply of BH_4 to the iNOS enzyme.

L-arginine availability may also represent a mechanism by which NO production by iNOS may be governed. One important process in this regard is the transport of L-arginine across the plasma membrane. In various cell types, including macrophages and endothelial cells, L-arginine transport is up-regulated in response to inflammatory mediators which induce the expression of iNOS [55]. In macrophages, this enhanced L-arginine uptake is thought to be due to the induction of the MCAT-2B cationic amino acid transporter [56]. Interestingly, L-lysine and L-ornithine are also substrates for this transporter and are thereby able to inhibit L-arginine uptake competitively [55]. This mechanism for regulating NO production by iNOS is particularly important in the case of macrophages and other immune cells. Here, NO production by iNOS is solely dependent on the extracellular arginine concentration and consequently relies entirely on the uptake of arginine to satisfy substrate requirements for NOS [57]. NO production in these cells is therefore highly sensitive to inhibitors of arginine transport. In contrast, there is also evidence to suggest that in some cell types increased arginine synthesis can augment NOS substrate availability. This may be the result of a co-induction of iNOS and arginosuccinate synthase, the rate-limiting enzyme in arginine formation [58].

NO-dependent mechanisms involved in inflammation

Initially, an inflammatory response involves the release of 'acute phase' mediators and the recruitment of polymorphonuclear cells to the site of trauma. Following this process, which usually spans 4–6 hours, continued insult results in a chronic inflammatory response which is characterised by the activation of mononuclear cells and the expression of iNOS in a wide variety of cell types, including the macrophages themselves. Indeed, this profile of activity has been demonstrated by following the expression of iNOS during acute and chronic inflammation [59]. Constitutive NOS accounts for the majority of NO activity in the acute phase of inflammation; at this point, some polymorphonuclear cells express iNOS and a few resident macrophages also show positive staining for the inducible isoform. At the peak of chronic inflammation there is an eight-fold increase in total NOS activity, of which greater than 90% can be attributed to iNOS activity in activated macrophages. Activity is substantially reduced after 14 days as the inflammation disappears. Interestingly, the cytokines produced in the acute phase are predominantly IL-1β and TNF-α, both of which induce iNOS expression in a number of cell types. Conversely, as the inflammation subsides, the predominant cytokine is TGF-β, which tends to down-regulate the expression of iNOS. Thus, the activity of iNOS in the inflammatory process is tightly regulated. Cells that express iNOS subsequently produce large quantities of NO which is free to diffuse to adjacent cells to react with various target molecules, of which the best characterised are oxygen (and oxygen-derived radicals), thiols, and transition metals, particularly those forming the prosthetic groups in different enzymes resulting in modulation of activity. These actions are a prerequisite for the inflammatory actions of NO and are detailed below.

Inflammatory cell function

The potential involvement of NO in inflammation was probably first highlighted by its role as an effector molecule of macrophage activity. Elevations in urinary nitrate can be observed in patients with fever and diarrhoea as well as in rats exposed to bacterial LPS [60]. Subsequently, it was demonstrated that the activated macrophage was the major source of this nitrate [61]. Moreover, the cytotoxic actions of such macrophages were shown to be L-arginine-dependent and inhibited by N^G-monomethyl-L-arginine (L-NMMA) [62–64], suggesting the involvement of NO. The release of macrophage-derived nitrite was not observed in freshly isolated cells, but required incubation for at least 6 hours with LPS; this phenomenon was explained by the requirement for *de novo* (NOS) protein synthesis [65, 66]. In addition to macrophages, NO is also released by mast cells [67] and neutrophils [68, 69]. Indeed, the anti-inflammatory properties of glucocorticoids may be related, at least in part, to their inhibition of iNOS induction, as can be demonstrated in immune cells [69].

Neutrophils and macrophages are the predominant immune cells at sites of acute and chronic inflammation, respectively. Human neutrophils synthesise NO at levels which can cause vasodilatation and inhibit thrombin-induced platelet aggregation [70–72]. It appears that neutrophil-derived NO is important in certain models of inflammation and it has been reported to possess antimicrobial activity [73, 74]. However, the concentrations of nitrite/nitrate produced by neutrophils are substantially lower than that produced by macrophages [75] and, consequently, it is unclear whether neutrophils produce NO in quantities sufficient to exert a pathological effect; hence, it appears that macrophages are the most important immune cell in this respect. Characterisation of the expression of iNOS in rodent macrophages by various cytokines and transcription factors has been comprehensive (see above), yet induction of iNOS in human macrophages remains controversial. In addition to neutrophils and macrophages numerous other cell types are capable of expressing iNOS, depending upon the appropriate stimulation, including the vascular endothelium, respiratory epithelium, keratinocytes, chondrocytes, fibroblasts, myocytes and astrocytes [76]. The function of iNOS in such cells, and its relation to inflammation, sepsis and host defence remains largely to be explored.

NO also has an important regulatory role on immune cell function in addition to acting as an effector molecule. Indeed, NO can both up- and down-regulate the functions of immune cells, depending on their subset and the local concentration of NO. Furthermore, this regulatory activity correlates well with the ability of NO to both up- and down-regulate NF-κB activity (as described above) and may indicate that this transcription factor is important in controlling immune cell function. NO, when applied exogenously or released from activated macrophages, can inhibit T-cell proliferation [77, 78]. In a similar fashion, the large quantities of NO often released by tumours may account, at least in part, for the immunosuppression observed in patients with cancer [79]. Probably the best indication as to the regulation of T-cell function by NO is provided by murine models of malaria infection. Spleen cells from mice infected with *Plasmodium chabaudi chabaudi* produce significant amounts of IFN-γ and IL-2 when cultured with malarial antigens. In the presence of L-NMMA, the release of these cytokines is significantly enhanced, suggesting that NO is dampening the activity of these cells [80]. Conversely, the addition of the NO-donor S-nitroso-N-acetylpenicillamine (SNAP) markedly inhibits the synthesis of IFN-γ and IL-2. Moreover, the inhibitory effect of NO in these cells can be reversed by the addition of exogenous IL-2 (but not IFN-γ), suggesting that mechanistically NO prevents the synthesis and/or secretion of IL-2 and thus exerts a widespread regulatory action on T-cells (since IL-2 is an autocrine regulator of T-cell activity). A further mechanism by which NO can govern the immune process is to down-regulate the expression of MHC class II molecules on the surface of antigen-presenting cells [81].

NO can also act in an anti-inflammatory manner, in direct contrast to the action described above. For instance, mice treated with L-NMMA have reduced local inflammatory responses following carrageenin injection into the paw [82]. This phenomenon is accompanied by a reduction in Th_1-cell responses, as assessed by the production of IL-2 and IFN-γ. In a collagen-induced arthritis model, treatment of rats with N^G-nitro-L-arginine methyl ester (L-NAME) dramatically reduces the severity of the inflammation but significantly enhances T-cell proliferation [83].

The ability of NO to both up- and down-regulate T-cell function and the inflammatory process in general may be linked to the capacity of NO to regulate T-cell activity differentially. Thus, when T-cells specific for malarial antigens are challenged, only the Th_1 population responds by synthesising NO, whereas Th_2 cells do not [80]. Moreover, the production of IL-4 by Th_2 cells is relatively resistant to the effects of NO (i.e. NO-donors) when compared to IL-2 and IFN-γ production by Th_1 cells. This 'feedback loop' has very important implications for the regulation of host defence. The activity of Th_1 cells is generally associated with immunopathological processes, and therefore the production of NO by Th_1 cells (and activated macrophages) may be important in eradicating invading microorganisms. However, prolonged/excessive production of NO may result in self-damage and the ability of NO to curb the activity of Th_1 cells is likely to be an endogenous braking mechanism by which Th_1 cell over-activity is prevented. Furthermore, the inability of NO to modulate Th_2 cell function provides a second process which can operate to provide exquisite control of the inflammatory process.

Haemodynamics

The interaction of NO with iron within the haem moiety of soluble guanylate cyclase is the best-studied intracellular target site for NO and mediates the haemodynamic effects of NO, both physiologically and during inflammation and sepsis. Indeed, the profound vasodilator activity of NO during pathophysiological episodes is thought to represent an adaptive response to offset the increased concentrations of circulating constrictors and to allow increased tissue perfusion and the delivery of inflammatory cells and mediators. Unfortunately, during sepsis the large amounts of NO released by iNOS often result in a severe systemic hypotension; this represents a life-threatening condition in which multiple organ failure and death are possible outcomes. Nevertheless, this may be prevented by specific iNOS inhibitors, which can reduce the generation of NO. Indeed, limited studies in humans using a non-selective NOS inhibitor, L-NMMA, suggest that such compounds may be effective [84]. A problem still exists, however. As described below, NO has important roles to play in mediating the cytostatic/cytocidal actions of inflammatory cells and in the management of haemodynamics in sepsis. Hence, if one is to remove the production of NO by iNOS in such conditions, it may reverse the vasodilator actions of NO, but it will also interfere with host defence. As such, a selective guanylate cyclase inhibitor, which can differentiate between the haemodynamic and inflammatory properties of NO may be a preferable choice of treatment.

The role of iNOS-synthesised NO in the hypotension associated with septic shock has been proven by numerous *in vivo* models, as has the use of NOS inhibitors, particularly the recent development of iNOS-selective compounds. In a rodent model of gram-positive shock, dexamethasone (an inhibitor of iNOS expression) and aminoguanidine (a selective iNOS inhibitor) prevent the drop in mean arterial blood pressure and pO_2, and the increase in plasma glutamate-pyruvate transaminase (a marker of hepatocellular injury), urea/creatinine (markers of renal dysfunction) and nitrate (a metabolite of NO). Furthermore, the effects of dexamethasone are due to decreased transcription and expression of iNOS, as measured in the aorta, liver and lung [85]. Selective iNOS inhibitors also attenuate the microvascular injury (as assessed by vascular leakage of serum albumin) associated with the early pathophysiological changes following LPS administration [86]. Similar observations have been found with further, more selective inhibitors of iNOS, including 1-amino-2-hydroxy-guanidine and aminoethyl-iso-thiourea [87]. The effects of such compounds have a dramatic effect on survival of animals exposed to endotoxin [88], which explains the interest in the development of these iNOS inhibitors. Recently, the identification of a family of selective tyro-sine kinase inhibitors, the tyrphostins, has revealed a further class of compound which can reduce iNOS expression during sepsis and thereby reverse the haemo-dynamic problems associated with this pathology. In this regard, tyrphostin AG 126 has been shown to protect mice against LPS-induced lethal toxicity, an effect which correlates with a reduction in TNF-α and NO production [89].

Regulation of protein synthesis and enzyme activity

An additional mechanism by which NO modulates inflammatory processes is via regulation of protein synthesis. In cultured hepatocytes stimulated with cytokines and endotoxin there is a NO-dependent depression of protein synthesis [90]. Exoge-nous administration of NO *in vitro* mimics this phenomenon [91]. Moreover, NO has also been shown to prevent TGF-β and collagen synthesis in mesangial cells [92] and may be involved in regulation of matrix protein synthesis. Contrastingly, NO may also be involved in enhancing the synthesis of certain proteins. This appears particularly important in the ability of NO to activate the production of eico-sanoids, which are obligatory to an inflammatory response. NO is able to activate directly cyclooxygenase (COX), the rate-limiting enzyme in prostanoid synthesis, in astrocytes, endothelial cells and macrophages [93–95]. Interestingly, NO appears to be able to activate both COX I and COX II, the latter being induced during inflam-matory episodes by the influence of similar factors (i.e. cytokines and bacterial prod-ucts) to those which regulate iNOS expression; in this way, NO and prostanoids synergise in regulating the inflammatory process. The mechanism involved is unclear, but the avid interaction of NO with haem moieties, as possessed by COX isozymes, may result in enzyme activation. In addition, COX-II expression, like iNOS, is regulated by the transcription factor, NF-κB; this therefore provides an alternate mechanism by which NO can alter eicosanoid synthesis.

Cytotoxicity

NO inhibits the activity of many enzymes and this appears to be extremely impor-tant in mediating the cytotoxic and cytostatic actions of NO, necessary for the destruction and eradication of pathogens. These actions of NO are exemplified in its antimicrobial activity; NO has powerful cytotoxic actions against many microbes, including *Cryptococcus neoformans*, *Schistosoma mansoni*, *Trypanoso-*

ma cruzi, *Plasmodium falciparum* and *Mycobacterium tuberculosis* [96]. Furthermore, NO is important in the killing of parasites such as *Leishmania major* [97].

Several of the enzymes inhibited by NO are located in the mitochondria and are involved in the electron transport chain; thus inhibition of respiration may well represent an important cytotoxic action of NO. NO has been shown to react with the iron-sulphur clusters in aconitase (tricarboxylic acid cycle and iron-regulation) and complexes I, II and cytochrome c oxidase of the respiratory chain. The end result of such actions is to deplete the target cells of ATP and hence cause cytotoxicity. In addition, inhibition of the respiratory chain results in a build-up of electrons in the system which can then reduce molecular oxygen to produce $O_2^{\cdot-}$, which is potentially cytotoxic due to subsequent formation of H_2O_2. Furthermore, since NO is a radical, it reacts at near diffusion controlled rates with other radicals; this is exemplified by the interaction of NO with $O_2^{\cdot-}$ yielding peroxynitrite ($ONOO^-$). Indeed, this reaction proceeds 3 times more rapidly than the disproportionation of $O_2^{\cdot-}$ by SOD. The combination of NO and $O_2^{\cdot-}$ to give $ONOO^-$ has received considerable attention over the past few years since it has been suggested that $ONOO^-$ may represent an important mediator of cytotoxicity and cytostasis [98]. The mechanism of $ONOO^-$-induced toxicity is unclear, but may involve the nitrosation and/or nitration of tyrosine and thiol-containing residues on various proteins; such modifications have been suggested to alter protein function and thereby disrupt cellular activity. The cytotoxic potential of $ONOO^-$ is illustrated in its irreversible inhibition of enzymes in the electron transport chain [99].

Additional mechanisms by which NO can disrupt cellular function include impairment of DNA synthesis by inhibition of ribonucleotide reductase [100] and direct DNA deamination reactions [101]. NO can also inactivate phosphoenolpyruvate carboxykinase and glyceraldehyde-3-phosphate dehydrogenase, both of which are important in glucose metabolism, and thereby reduce the energy production of target cells [102]. Interestingly, immune cells are extremely efficient at providing themselves with energy via anaerobic metabolism. Therefore, they may actually prevent the NO they generate from being self-damaging by being able to function independently of the aerobic respiratory enzymes which NO can disrupt (above). If this were the case, cells which cannot switch between aerobic and anaerobic metabolism (e.g. tumor cells) would be more susceptible to the cytotoxic actions of NO.

Inflammatory pain

A major consequence of the release of 'acute phase' mediators during an inflammatory reaction is the subsequent generation of factors which modulate the threshold and sensitivity of nociceptors; hence the phenomenon of inflammatory pain. Many compounds are thought to act in this regard including COX metabolites, sympathomimetic amines, bradykinin and cytokines (released by infiltrating inflammatory cells). Such agents are believed to sensitise otherwise dormant nociceptors which are then triggered by previously ineffective stimuli. Consequently, drugs inhibiting COX (e.g. NSAIDS), or those antagonising sympathomimetic amines (e.g. β-blockers) or cytokines, prevent sensitisation of nociceptors and reduce inflammatory pain.

In accord with its role as an inflammatory mediator, NO is associated with the development of inflammatory pain. There is evidence to suggest that constitutive and inducible enzymes are involved, both centrally and peripherally. However, in an analogous fashion to its ambiguous role as an anti- or pro- inflammatory agent, it is unclear whether NO enhances or reduces the pain associated with an inflammatory response. Formalin (intraplantar)-induced nociceptive behaviour in the mouse is characterised by early and late phase responses which represent C-fibre activation and central sensitisation, respectively. L- but not D- arginine, when coadministered with formalin, enhances the second- but not first- phase nociceptive response (i.e. hindpaw licking) [103]. Accordingly, L-NAME (but not D-NAME) causes a dose-dependent, anti-nociceptive effect in the same model. These observations are echoed by the use of a centrally-acting nNOS inhibitor, 7-nitroindazole [104]. This compound produces a dose-dependent anti-nociceptive activity in the second phase response following intraplantar formalin. Likewise, formalin-induced nociceptive behaviour can be facilitated by the NO-donor NOC-18 when administered intracerebroventricularly during the second but not first phase response following formalin administration [105]. The focus of NO-mediated effects to the second phase of formalin-induced nociception has been confirmed by electrophysiological studies. In these studies, recordings from single dorsal root horn neurons are used to assess formalin-induced nociceptive behaviour and it has been demonstrated that i.v. administration of L-NAME prior to intraplantar formalin greatly reduces the prolonged second peak of firing with only a minor effect on the short-duration first peak [106]. In human studies involving the elicitation of pain from the hand vein, NO also appears to exert nociceptive effects. Following administration of bradykinin or hyperosmolar solutions to isolated perfused hand veins, NOS inhibitors markedly reduce the discomfort [107]. Moreover, NO itself can produce pronounced nociception [108]. Further, NO may be important in the synaptic plasticity following noxious peripheral stimulation. In nNOS knockout mice, the formalin-induced nociceptive behaviour remains intact but the neuroplasticity cannot be blocked by an NOS inhibitor, as is observed in wild-type animals. These results suggest that nNOS-derived NO may be important in nociception-induced neuroplasticity whereas an alternative isoform of NOS is responsible for the production of NO involved in the development of inflammatory pain [109].

Conversely, however, agents that can stimulate formation of cGMP (akin to those increasing cAMP levels) have been shown to cause functional nociceptor down-regulation [110]. This applies to agents activating sGC (e.g. nitrovasodilators) or to cGMP-phosphodiesterase (type V) inhibitors; agents blocking sGC activity abolish this anti-nociceptive effect. The ability of NO to reduce inflammatory pain may also underlie the anti-nociceptive effects of endogenous compounds. Opiates block inflammatory hyperalgesia [111, 112] and it has been demonstrated that peripherally-acting opiates that cause down-regulation of nociceptors activate the NO-cGMP signal transduction pathway in pain sensory neurons [110, 113].

In summary, the role of NO in inflammatory pain is as complex and uncertain as its role in the inflammatory response as a whole and the central and peripheral effects of NO will require careful dissection in order to elucidate whether NO is indeed nociceptive or anti-nociceptive.

Nitric oxide and other inflammatory disease states

In addition to those described above, NO has been implicated in many other inflammatory disease states and models, of which the following have been most extensively studied. Rheumatoid arthritis is a chronic autoimmune inflammatory disease characterised by, amongst other symptoms, injury and proliferation of the synovium lining joint cavities. Elevated concentrations of nitrite and nitrate have been found in the synovial fluid of arthritic patients, suggesting an increased local release of NO, and chondrocytes in culture express iNOS and decrease their proteoglycan synthesis when stimulated by inflammatory cytokines [114, 115]. Adjuvant-induced arthritis in rats is exacerbated by L-arginine and reduced by NOS inhibitors [116]. Thus, the elevated production of NO in rheumatoid arthritis, and its contribution to disease progression, seems clear. NO also appears important in the cell-mediated rejection of allogeneic transplants. Treatment with NOS inhibitors reduces mucosal pathology and epithelial lymphocyte infiltration in mice with intestinal graft-vs-host reaction; in addition, NOS inhibitors reduce the stimulated activity of cytotoxic T-cells, suggesting an effector role for NO in mediating tissue destruction in this disease state [117]. Moreover, electron paramagnetic resonance analysis has identified iron-nitrosyl complexes in tissues and blood from rejected allografts of rat hearts [118]. Finally, rats which acutely reject allografts, or with graft-vs-host reaction, have elevated levels of serum nitrite/nitrate, which can be decreased in animals by administration of immunosuppressive therapy. NO also appears to play a role in inflammation of the gastrointestinal tract (which is regulated physiologically by NO released by non-adrenergic, non-cholinergic nerves) [119] and in renal inflammation [120].

Summary

The increased understanding of the physiological and pathological roles of NO have demonstrated that this unique molecule plays an important role in the process of inflammation. This encompasses both NO derived from constitutive NOS isozymes,

45

which appears important in the early stages of an inflammatory response, through to high-output production of NO by the inducible NOS, which is fundamental to chronic inflammatory disease. The actions of NO in relation to inflammation are multi-faceted and as a consequence it remains unclear if NO is pro- or anti-inflammatory; ultimately it is likely to be both, very much dependent on the local concentrations of NO and the prevailing immunological status of the host. In the search for therapeutic agents which may modulate NO activity and hence inflammation in general, concentration and time of administration will be vital for effective treatment. This is particularly true with the recent data suggesting that NO can modulate directly the expression and activity of numerous pro-inflammatory proteins. Thus, not only does NO have direct actions which mediate certain aspects of inflammation, but it has far wider-reaching consequences for inflammation in general, indirectly, due to its regulation of transcription and protein activity. It is of fundamental importance therefore, that the role of NO in the process of inflammation is understood in its entirety, so that agents can be developed to modulate the actions of NO and provide relief from inflammatory disease.

References

1   Knowles RG, Moncada S (1994) Nitric oxide synthases in mammals. *Biochem J* 298: 249–258
2   Nathan C (1992) Nitric oxide as a secretory product of mammalian cells. *FASEB J* 6: 3051–3064
3   Djordjevi S, Roberts DL, Wang M, SheaT, Camitta MG, Masters BS, Kim JJ (1995) Crystallization and preliminary x-ray studies of NADPH-cytochrome P450 reductase. *Proc Natl Acad Sci USA* 92: 3214–3218
4   Bredt DS, Hwang PM, Glatt CE, Lowenstein C, Reed RR, Snyder SH (1991) Cloned and expressed nitric oxide synthase structurally resembles cytochrome P450 reductase. *Nature* 351: 714–718
5   Stuehr DJ, Ikeda-Saito M (1992) Spectral characterization of brain and macrophage nitric oxide synthases Cytochrome P450-like hemoproteins that contain a flavin semiquinone radical. *J Biol Chem* 267: 20547–20550
6   White KA, Marletta MA (1992) Nitric oxide synthase is a cytochrome P450-type hemoprotein. *Biochemistry* 31: 6627–6631
7   Wang J, Stuehr DJ, Ikeda-Saito M, Rousseau DL (1993) Heme coordination and structure of the catalytic site in nitric oxide synthase. *J Biol Chem* 268: 22255–22258
8   McMillan K, Masters BS (1995) Prokaryotic expression of the heme- and flavin-binding domains of rat neuronal nitric oxide synthase as distinct polypeptides: identification of the heme-binding proximal thiolate ligand as cysteine-415. *Biochemistry* 34: 3686–3693
9   Sheta EA, McMillan K, Masters BS (1994) Evidence for a bidomain structure of constitutive cerebellar nitric oxide synthase. *J Biol Chem* 269: 15147–15153

10   Stuehr DJ, Abu-Soud HM, Rousseau DL (1995) Control of electron transfer in neuronal nitric oxide synthase by calmodulin, substrate analogs, and nitric oxide. *Adv Pharmacol* 34: 207–213

11   Garcia-Cardena G, Oh P, Liu J, Schnitzer JE, Sessa WC (1996) Targeting of nitric oxide synthase to endothelial caveolae via palmitoylation: implications for nitric oxide signaling. *Proc Natl Acad Sci USA* 93: 6448–6453

12   Stuehr DJ, Cho HJ, Kwon NS, Weise MF, Nathan CF (1991) Purification and characterization of the cytokine-induced macrophage nitric oxide synthase; an FAD- and FMN- containing flavoprotein. *Proc Natl Acad Sci USA* 86: 7773–7777

13   Cho AK, Lindeke B (1988) *Biotransformation of Organic Nitrogen Compounds*. Karger, Basel

14   Fukuto JM, Wallace GC, Hszeih R, Chaudhuri G (1992) Chemical oxidation of N-hydroxyguanidine compounds. *Biochem Pharmacol* 43: 607–613

15   Fukuto JM, Stuehr DJ, Feldman PL, Bova MP, Wong, P (1993) Peracid oxidation of an N-hydroxyguanidine compound: a chemical model for the oxidation of N-hydroxy-L-arginine by nitric oxide synthase. *J Med Chem* 36: 2666–2670

16   Hobbs AJ, Fukuto JM, Ignarro LJ (1994) Formation of free nitric oxide from L-arginine by nitric oxide synthase: direct enhancement of generation by superoxide dismutase. *Proc Natl Acad Sci USA* 91: 10992–10996

17   Schmidt HHHW, Hofmann H, Schindler U, Shutenko ZS, Cunningham DD, Feelisch M (1996) No NO from NO synthase. *Proc Natl Acad Sci USA* 93: 14492–14497

18   Fujii E, Irie K, Uchida Y, Tsukahara F, Muraki T (1994) Possible role of nitric oxide in 5-hydroxytryptamine-induced increase in vascular permeability in mouse skin. *Naunyn-Schmiedebergs Arch Pharmacol* : 350: 361–364

19   Kajekar R, Moore PK, Brain SD (1995) Essential role for nitric oxide in neurogenic inflammation in rat cutaneous microcirculation Evidence for an endothelium-independent mechanism. *Circ Res* 76: 441–447

20   Moncada S, Palmer RMJ, Higgs EA (1990) Relationship between prostacyclin and nitric oxide in the thrombotic process. *Thromb Res Suppl* 11: 3–13

21   Kubes P, Suzuki M, Granger DM (1991) Nitric oxide: An endogenous modulator of leukocyte adhesion. *Proc Natl Acad Sci USA* 88: 4651–4655

22   Chartrain NA, Geller DA, Koty PP, Sitrin NF, Nussler AK, Hoffman EP, Billiar TR, Hutchinson NI, Mudgett JS (1994) Molecular cloning, structure, and chromosomal localization of the human inducible nitric oxide synthase gene. *J Biol Chem* 269: 6765–6772

23   Kaling M, Kugler W, Ross K, Zoidl C, Ryffel GU (1991) Liver-specific gene expression: A-activator-binding site, a promoter module present in vitellogenin and acute phase genes. *Mol Cell Biol* 11: 93–101

24   Xie QW, Kashiwabara Y, Nathan C (1994) Role of transcription factor NF-kappaB/Rel in induction of nitric oxide synthase. *J Biol Chem* 269: 4705–4708

25   Martin E, Nathan C, Xie QW (1994) Role of interferon regulatory factor 1 in induction of nitric oxide synthase. *J Exp Med* 180: 977–984

26   Lowenstein CJ, Alley EW, Raval P, Snowman AM, Snyder SH, Russell SW, Murphy WJ (1993) Macrophage nitric oxide synthase gene: two upstream regions mediate induction by interferon gamma and lipopolysaccharide. *Proc Natl Acad Sci USA* 90: 9730–9734

27   De Vera ME, Shapiro RA, Nussler AK, Mudgett JS, Simmons RL, Morris SM, Billiar TR, Geller DA (1995) Transcriptional regulation of human inducible nitric oxide synthase (NOS2) gene by cytokines: Initial analysis of the human NOS2 promoter. *Proc Natl Acad Sci USA* 93: 1054–1059

28   Weinberg JB, Misukonis MA, Shami PJ, Mason SN, Sauls DL, Dittman WA, Wood ER, Smith GK, McDonald B, Bachus KE (1995) Human mononuclear phagocytic inducible nitric oxide synthase (iNOS): Analysis of iNOS mRNA, iNOS protein, biopterin, and nitric oxide production by blood monocytes and peritoneal macrophages. *Blood* 86: 1184–1195

29   Lenardo MJ, Baltimore D (1989) NF-κB: A pleiotropic mediator of inducible and tissue-specific gene control. *Cell* 58: 227–229

30   Baeuerle PA (1991) The inducible transcription factor NF-κB: regulation by distinct protein subunits. *Biochim Biophys Acta* 1072: 63–80

31   Baeuerle PA, Baltimore D(1988) IκB: A specific inhibitor of the NF-κB transcription factor. *Science* 242: 540–546

32   Urban MB, Schreck R, Baeuerle PA (1991) NF-κB contacts DNA by a heterodimer of the p50 and p65 subunit. *EMBO J* 10: 1817–1825

33   Palombella VJ, Rando OJ, Goldberg AL, Maniatis T (1994) The ubiquitin-proteasome pathway is required for processing the NF-κB1 precursor protein and the activation of NF-κB. *Cell* 78: 773–785

34   Traenckner EB, Pahl HL, Henkel T, Schmidt KN, WilkS, Baeuerle PA (1995) Phosphorylation of human IκB-α on serines 32 and 36 controls IκB-α proteolysis and NF-κB activation in response to diverse stimuli. *EMBO J* 14: 2876–2883

35   Henkel T, Machleidt T, Alkalay I, Kronke M, Ben-Neriah Y, Baeuerle, PA (1993) Rapid proteolysis of IκB-α in response to phorbol ester, cytokines and lipopolysaccharide is a necessary step in the activation of NF-κB. *Nature* 365: 182–185

36   Schreck R, Albermann K, Baeuerle PA (1992) NF-κB: an oxidative stress-response transcription factor of eukaryotic cells. *Free Rad Res Commun* 17: 221–237

37   Schreck R, Reiber P, Baeuerle PA (1991) Reactive oxygen intermediates as apparently widely used messengers in the activation of NF-κB transcription factor and HIV-1. *EMBO J* 10: 2247–2258

38   Hibbs JB, Taintor RR, Vavrin Z, Rachlin EM (1988) Nitric oxide: a cytotoxic activated macrophage effector molecule. *Biochem Biophys Res Commun* 157: 87–94

39   Sherman MP, Aeberhard EE, Wong VZ, Griscavage JM, Ignarro LJ (1993) Pyrrolidine dithiocarbamate inhibits induction of nitric oxide synthase activity in rat alveolar macrophages. *Biochem Biophys Res Commun* 191: 1301–1308

40   Mulsch A, Schray-Utz B, Mordvintcev PI Hauschildt, S, Busse R (1993) Diethyldithiocarbamate inhibits induction of macrophage NO synthase. *FEBS Lett* 321: 215–218

41   Griscavage JM, Wilk S, Ignarro LJ (1995) Serine and cysteine proteinase inhibitors pre-

vent nitric oxide production by activated macrophages by interfering with transcription of the inducible NO synthase gene. *Biochem Biophys Res Commun* 215: 721–729

42 Griscavage JM, Wilk S, Ignarro LJ (1996) Inhibitors of the proteasome pathway interfere with induction of nitric oxide synthase in macrophages by blocking activation of nuclear factor kappa-B. *Proc Natl Acad Sci USA* 93: 3308–3312

43 Radomski MW, Palmer RM, Moncada S (1990) Glucocorticoids inhibit the expression of an inducible, but not the constitutive, nitric oxide synthase in vascular endothelial cells. *Proc Natl Acad Sci USA* 87: 10043–10047

44 Di Rosa M, Radomski MW, Carnuccio R, Moncada S (1990) Glucocorticoids inhibit the induction of nitric oxide synthase in macrophages. *Biochem Biophys Res Commun* 171: 1246–1252

45 Ray A, Prefontaine KE (1994) Physical association and functional antagonism between the p65 subunit of transcription factor NF-κB and the glucocorticoid receptor. *Proc Natl Acad Sci USA* 91: 752–756

46 Kleinert H, Euchenhofer C, Ihrig-Biedert I, Forstermann U (1996) Glucocorticoids inhibit the induction of nitric oxide synthase II by down-regulating cytokine-induced activity of transcription factor nuclear factor-κB. *Mol Pharmacol* 49: 15–21

47 Auphan N, DiDonato JA, Rosette C, Helmberg A, Karin M (1995) Immunosuppression by glucocorticoids: inhibition of NF-κB activity through induction of IκB synthesis. *Science* 270: 286–290

48 Aeberhard EE, Henderson SA, Arabolos NS, Griscavage JM, Castro FE, Barrett CT, Ignarro LJ (1995) Nonsteroidal anti-inflammatory drugs inhibit expression of the inducible nitric oxide synthase gene. *Biochem Biophys Res Commun* 208: 1053–1059

49 Kopp E, Ghosh S (1994) Inhibition of NF-κB by sodium salicylate and aspirin. *Science* 265: 956–959

50 Peng HB, Libby P, Liao JK (1995) Induction and stabilization of I kappa B alpha by nitric oxide mediates inhibition of NF-kappa B. *J Biol Chem* 270: 14214–14219

51 Matthews JR, Botting CH, Panico M, Morris HR, Hay RT (1996) Inhibition of NF-kappa B DNA binding by nitric oxide. *Nucl Acids Res* 24: 2236–2242

52 Di-Silvio M, Geller DA, Gross SS, Nussler A, Freeswick P, Simmons RL, Billiar TR (1993) Inducible nitric oxide synthase activity in hepatocytes is dependent on the coinduction of tetrahydrobiopterin synthesis. *Adv Exp Med Biol* 338: 305–308

53 Gross SS, Levi R, Madera A, Park KH, Vane JR, Hattori Y (1993) Tetrahydrobiopterin synthesis is induced by LPS in vascular smooth muscle and is rate-limiting for nitric oxide production. *Adv Exp Med Biol* 338: 295–300

54 Gross SS, Levi R (1992) Tetrahydrobiopterin synthesis. An absolute requirement for cytokine-induced nitric oxide synthase generation by vascular smooth muscle. *J Biol Chem* 267: 25722–25729

55 Bogle RG, Baydoun AR, Pearson JD, Moncada S, Mann GE (1992) L-Arginine transport is increased in macrophages generating nitric oxide. *Biochem J* 284: 15–18

56 Closs EI, Lyons CR, Kelly C, Cunningham JM (1993) Characterization of the third member of the MCAT family of cationic amino acid transporters. Identification of a

domain that determines the transport properties of the MCAT proteins. *J Biol Chem* 268: 20796–20800

57 Assreuy J, Moncada S (1992) A perfusion system for the long term study of macrophage activation. *Br J Pharmacol* 107: 317–321

58 Nussler AK, Billiar TR, Liu Z-Z, Morris SM (1994) Coinduction of nitric oxide synthase and arginosuccinate synthetase in a murine cell line. Implications for regulation of nitric oxide production. *J Biol Chem* 269: 1257–1261

59 Vane JR, Mitchell JA, Appleton I, Tomlinson A, Bishop-Bailey D, Croxtall J, Willoughby DA (1994) Inducible isoforms of cyclooxygenase and nitric oxide synthase in inflammation. *Proc Natl Acad Sci USA* 91: 2046–2050

60 Wagner DA, Young VR, Tannenbaum SR (1983) Mammalian nitrate biosynthesis: incorporation of $^{15}NH_3$ into nitrate is enhanced by endotoxin treatment. *Proc Natl Acad Sci USA* 80: 4518–4521

61 Stuehr DJ, Marletta MA (1987) Synthesis of nitrite and nitrate in murine macrophage cell lines. *Cancer Res* 47: 5590–5594

62 Hibbs JB, Taintor RR, Vavrin Z Macrophage cytotoxicity: role for L-arginine deiminase activity and imino nitrogen oxidation to nitrite. *Science* 1987; 235: 473–476

63 Hibbs JB, Taintor RR, Vavrin Z, Rachlin EM (1988) Nitric oxide: a cytotoxic activated macrophage effector molecule. *Biochem Biophys Res Commun* 157: 87–94

64 Stuehr DJ, Gross SS, Sakuma I, Levi R, Nathan C (1989) Activated murine macrophages secrete a metabolite of arginine with the bioactivity of endothelium-derived relaxing factor and the chemical reactivity of nitric oxide. *J Exp Med* 169: 1011–1020

65 Marletta MA, Yoon PS, Iyengar R, Leaf CD, Wishnok JS (1988) Macrophage oxidation of L-arginine to nitrite and nitrate: nitric oxide is an intermediate. *Biochemistry* 27: 8706–8711

66 Tayeh MA, Marletta MA (1989) Macrophage oxidation of L-arginine to nitric oxide, nitrite and nitrate Tetrahydrobiopterin is required as a co-factor. *J Biol Chem* 264: 19654–19658

67 Salvemini D, Masini E, Anggard E, Mannaioni PF, Vane JR Synthesis of a nitric oxide-like factor from L-arginine by rat serosal mast cells: stimulation of guanylate cyclase and inhibition of platelet aggregation. *Biochem Biophys Res Commun* 1990; 169: 596–601

68 McCall TB, Feelisch M, Palmer RMJ, Moncada S (1991) Identification of N-iminoethyl-L-ornithine as an irreversible inhibitor of nitric oxide synthase in phagocytic cells. *Br J Pharmacol* 102: 234–238

69 McCall TB, Palmer RMJ, Moncada S (1991) Induction of nitric oxide synthase in rat peritoneal macrophages and its inhibition by dexamethasone. *Eur J Immunol* 21: 2523–2527

70 McCall TB, Boughton-Smith NK, Palmer RMJ, Moncada S (1989) Synthesis of nitric oxide from L-arginine by neutrophils Release and interaction with superoxide anions. *Biochem J* 261: 293–296

71 Schmidt HHHW, Seifert R, Bohme E (1989) Formation and release of nitric oxide from

human neutrophils and HL-60 cells induced by a chemotactic peptide, platelet activating factor and leukotriene B$_4$. *FEBS Lett* 244: 357–360

72 Salvemini D, Misko TP, Masferrer JL, Seibert K, Currie MG, Vane JR (1989) Human neutrophils and mononuclear cells inhibit platelet aggregation by releasing a nitric oxide-like factor. *Proc Natl Acad Sci USA* 90: 7240–7244

73 Albina JE, Mills CD, Henry WL, Caldwell MD (1990) Temporal expression of different pathways of L-arginine metabolism in healing wounds. *J Immunol* 144: 3877–3880

74 Malawista SE, Montgomery RR, Van Blaricom G (1992) Evidence for reactive nitrogen intermediates in killing of staphylococci by human neutrophil cytoplasts. A new microbicidal pathway for polymorphonuclear leukocytes. *J Clin Invest* 90: 631–635

75 Padgett EL, Pruett SB (1995) Rat, mouse and human neutrophils stimulated by a variety of activating agents produce much less nitrite than rodent macrophages. *Immunology* 84: 135–141

76 Nathan C, Xie QW (1994) Nitric oxide synthases: Roles, tolls, and controls. *Cell* 78: 915–918

77 Hoffman RA, Langrehr JM, Billiar TR, Curran RD, Simmons RL (1990) Alloantigen-induced activation of rat splenocytes is regulated by the oxidative metabolism of L-arginine. *J Immunol* 145: 2220–2226

78 Schleifer KW, Mansfield JM (1993) Suppressor macrophages in African trypanosomiasis inhibit T cell proliferative response by nitric oxide and prostaglandins. *J Immunol* 151: 5492–5503

79 Lejeune P, Lagadec P, Onier N, Pinard D, Oshima H, Jeannin JF (1994) Nitric oxide involvement in tumor-induced immunosuppression. *J Immunol* 152: 5077–5083

80 Taylor-Robinson AW, Liew FY, Severn A, Xu D, McSorely S, Garside P, Padron J, Phillips RS (1994) Regulation of the immune response by nitric oxide differentially produced by T-helper type-1 and T-helper type-2 cells. *Eur J Immunol* 24: 980–984

81 Sicher SC, Vazquez MA, Lu CY (1994) Inhibition of macrophage Ia expression by nitric oxide. *J Immunol* 153: 1293–1300

82 Ianaro A, O'Donnell CA, Di Rosa M, Liew FY (1994) A nitric oxide synthase inhibitor reduces inflammation, down-regulates inflammatory cytokines and enhances IL-10 production in carrageenin-induced oedema in mice. *Eur J Immunol* 82: 370–375

83 Ialenti A, Moncada S, Liew FY (1993) Modulation of adjuvant arthritis by endogenous nitric oxide. *Br J Pharmacol* 110: 701–706

84 Petros A, Lamb G, Leone A, Moncada S, Bennett D, Vallance P (1994) Effects of a nitric oxide synthase inhibitor in humans with septic shock. *Cardiovasc Res* 28: 34–39

85 Kengatharan KM, De Kimpe SJ, Thiemermann C (1996) Role of nitric oxide in the circulatory failure and organ injury in a rodent model of gram-positive shock. *Br J Pharmacol* 119: 1411–1421

86 Laszlo F, Whittle BJ, Evans SM, Moncada S (1995) Association of microvascular leakage with induction of nitric oxide synthase: effects of nitric oxide synthase inhibitors in various organs. *Eur J Pharmacol* 283: 47–53

87 Ruetten H, Southan GJ, Abate A, Thiemermann C (1996) Attenuation of endotoxin-

induced multiple organ dysfunction by 1-amino-2-hydroxyguanidine, a potent inhibitor of inducible nitric oxide synthase. *Br J Pharmacol* 118: 261–270

88   Wu CC, Chen SJ, Szabo C, Thiemermann C, Vane JR (1995) Aminoguanidine attenuates the delayed circulatory failure and improves survival in rodent models of endotoxic shock. *Br J Pharmacol* 114: 1666–1672

89   Novogrodsky A, Vanichkin A, Patya M, Gazit A, Osherov N, Levitzki A (1994) Prevention of lipopolysaccharide-induced lethal toxicity by tyrosine kinase inhibitors. *Science* 264: 1319–1322

90   Curran RD, Billiar TR, Stuehr DJ, Ochoa JB, Harbrecht BG, Flint SG, Simmons RL (1990) Multiple cytokines are required to induce hepatocyte nitric oxide production and inhibit total protein synthesis. *Ann Surg* 212: 462–471

91   Curran RD, Ferrari FK, Kispert PH, Stadler J, Stuehr DJ, Simmons RL, Billiar TR (1991) Nitric oxide and nitric oxide-generating compounds inhibit hepatocyte protein synthesis. *FASEB J* 5: 2085–2092

92   Craven PA, Studer RK, Felder J, Phillips S, De Rubertis FR (1996) Nitric oxide inhibition of transforming growth factor-beta and collagen synthesis in mesangial cells. *Diabetes* 46: 671–681

93   Molina-Holgado F, lledo A, Guaza C (1995) Evidence for cyclooxygenase activation by nitric oxide in astrocytes. *Glia* 15: 167–172

94   Davidge ST, Baker PN, Laughlin MK, Roberts JM (1995) Nitric oxide produced by endothelial cells increases production of eicosanoids through activation of prostaglandin H synthase. *Circ Res* 77: 274–283

95   Salvemini D, Misko TP, Masferrer JL, Seibert K, Currie MG, Needleman P (1993) Nitric oxide activates cyclooxygenase enzymes. *Proc Natl Acad Sci USA* 90: 7240–7244

96   Nussler AK, Billiar, TR (1993) Inflammation immunoregulation, and inducible nitric oxide synthase. *J Leukocyte Biol* 54: 171–178

97   Vouldoukis I, Riveros-Moreno V, Dugas B, Ouaaz F, Becherel P, Debre P, Moncada S, Mossalayi MD (1995) The killing of *Leishmania major* by human macrophages is mediated by nitric oxide induced after ligation of the Fc-ε RII/CD23 surface antigen. *Proc Natl Acad Sci USA* 92: 7804–7808

98   Radi R, Beckman JS, Bush KM, Freeman BA (1991) Peroxynitrite-induced membrane lipid peroxidation: the cytotoxic potential of superoxide and nitric oxide. *Arch Biochem Biophys* 288: 481–487

99   Lizasoain I, Moro MA, Knowles RG, Darley-Usmar V, Moncada S (1996) Nitric oxide and peroxynitrite exert distinct effects on mitochondrial respiration which are differentially blocked by glutathione or glucose. *Biochem J* 314: 877–880

100   Kwon NS, Stuehr DJ, Nathan CF (1991) Inhibition of tumor cell ribonucleotide reductase by macrophage-derived nitric oxide. *J Exp Med* 174: 761–767

101   Nguyen T, Brunson D, Crespi CL, Penman BW, Wishnok JS, Tannenbaum SR (1992) DNA damage and mutation in human cells exposed to nitric oxide *in vitro*. *Proc Natl Acad Sci USA* 89: 3030–3034

102   Ou J, Molina L, Kim YM, Billiar TR (1996) Excessive NO production does not account

for the inhibition of hepatic gluconeogenesis in endotoxemia. *Am J Physiol* 271: G621–G628

103 Kawabata A, Manabe S, Manabe Y, Takagi H (1994) Effect of topical administration of L-arginine on formalin-induced nociception in the mouse: a dual role of peripherally formed NO in pain modulation. *Br J Pharmacol* 112: 547–550

104 Moore PK, Babbedge RC, Wallace P, Gaffen ZA, Hart SL (1993) 7-nitro indazole, an inhibitor of nitric oxide synthase, exhibits anti-nociceptive activity in the mouse without increasing blood pressure. *Br J Pharmacol* 108: 296–297

105 Shibuta S, Mashimo T, Zhang P, Ohara A, Yoshiya IJ (1996) A new nitric oxide donor, NOC-18: exhibits a nociceptive effect in the rat formalin model. *Neurol Sci* 141: 1–5

106 Haley JE, Dickenson AH, Schachter M (1992) Electrophysiological evidence for a role of nitric oxide in prolonged chemical nociception in the rat. *Neuropharmacology* 31: 251–258

107 Kindgen-Milles D, Arndt JO (1996) Nitric oxide as a chemical link in the generation of pain from veins in humans. *Pain* 64: 139–142

108 Holthusen H, Arndt JO (1995) Nitric oxide evokes pain at nociceptors of the paravascular tissue and veins in humans. *J Physiol* 487: 253–258

109 Crosby G, Marota JJ, Huang PL (1995) Intact nociception-induced neuroplasticity in transgenic mice deficient in neuronal nitric oxide synthase. *Neuroscience* 69: 1013–1017

110 Duarte IDG, Dos Santos IR, Lorenzetti BB, Ferreira SH (1992) Analgesia by direct antagonism of nociceptor sensitization involves the arginine-nitric oxide-cGMP pathway. *Eur J Pharmacol* 217: 225–227

111 Ferreira SH, Nakamura M (1979) Prostaglandin hyperalgesia: the peripheral analgesic activity of morphine, enkephalins and opioid antagonists. *Prostaglandins* 18: 191–200

112 Ferreira SH, Nakamura M (1979) Prostaglandin hyperalgesia: relevance of the peripheral effect for the analgesic activity of opioid antagonists. *Prostaglandins* 18: 201–208

113 Ferreira SH, Duarte ID, Lorenzetti BB (1991) The molecular mechanism of action of peripheral morphine analgesia: stimulation of the cGMP system via nitric oxide release. *Eur J Pharmacol* 201: 212–122

114 Farrell AJ, Blake DR, Palmer RMJ, Moncada S (1992) Increased concentrations of nitrite in synovial fluid and serum samples suggest increased nitric oxide synthesis in rheumatic diseases. *Ann Rheumatol Dis* 51: 1219–1222

115 Hauselmann HJ, Oppliger L, Michel BA, Stefanovic-Racic M, Evans CH (1994) Nitric oxide and proteoglycan synthesis by human articular chondrocytes in alginate culture. *FEBS Lett* 352: 361–364

116 Ialenti A, Ianaro A, Moncada S, Di Rosa M (1992) Modulation of acute inflammation by endogenous nitric oxide. *Eur J Pharmacol* 211: 177–182

117 Garside P, Hutton AK, Severn A, Liew FY, Mowat AM (1992) Nitric oxide mediates intestinal pathology in graft versus host disease. *Eur J Immunol* 22: 2141–2145

118 Lancaster JR, Langrehr JM, Bergonia HA, Mirase N, Simmons RL, Hoffman RA(1992)

EPR detection of heme and non heme iron containing protein nitrosylation by nitric oxide during rejection of rat heart allograft. *J Biol Chem* 267: 10994–10998

119 Sanders KM, Ward SM (1992) Nitric oxide as a mediator of nonadrenergic noncholinergic neurotransmission. *Am J Physiol* 262: G379–G392

120 Raij L, Shultz PJ (1993) Endothelium-derived relaxing factor, nitric oxide: Effects on and production by mesangial cells and the glomerulus. *J Am Soc Nephrol* 3: 1435–1441

Overview of HO-1 in inflammatory pathologies

Dean Willis

Department of Experimental Pathology, William Harvey Research Institute, St. Bartholomew's and The Royal London School of Medicine, Charterhouse Square, London EC1M 6BQ, UK

Introduction

Heme consists of a protoporphyrin ring with a tightly bound iron atom, which can exist in both a ferrous (Fe^{2+}) and ferric (Fe^{3+}) state, held in the centre of the molecule by four nitrogen atoms. This molecule is of fundamental importance in many biochemical pathways including oxygen transport by hemoglobin and electron transport in the respiratory chain. Heme is also a prosthetic group for numerous enzymes, which most notably in the context of this book include cyclooxygenase (COX) and nitric oxide synthesis (NOS) [1]. Heme has been demonstrated to regulate protein synthase via elf-2a kinase [2]. Therefore, maintenance of the heme pool is of considerable importance in eukaryotic systems, with the levels of intracellular heme being controlled at both the level of biosynthesis and degradation.

δ-aminolevulinate synthase catalyses the first step in heme synthesis, the condensation of succinyl CoA and glycine to form δ-aminolevulinate, and is believed to be the rate limiting step in heme biosynthesis. Following formation of δ-aminolevulinate, progressive condensation, decarboxylation and oxidation steps follow to eventually form protoporphyrin IX. Finally, ferrous iron is inserted into the porphyrin ring via the catalytic activity of ferrochelatase to form the heme molecule [3]. These series of reactions take place both in the mitochondria and cytoplasm. In non-erythroid tissue, heme regulates its own production inhibiting the expression of the initial enzyme in its own biosynthetic cascade, δ-aminolevulinate synthase. In addition heme stimulates the expression of the rate limiting enzyme in heme degradation, heme oxygenase (HO). Catabolism of heme by HO results in the formation of free iron, carbon monoxide (CO) and the bile pigment biliverdin.

Traditionally, the biological significance of HO was believed to relate to its ability to recycle iron from heme, thus allowing reutilisation of the metal for heme synthesis. However, with the increasing importance of reactive oxygen and nitrogen species (ROS, RNS) in pathophysiological processes and the emerging biological actions of both carbon monoxide and the bile pigments, the role of HO in both normal and pathological conditions is being re-evaluated.

Inducible Enzymes in the Inflammatory Response, edited by D.A. Willoughby and A. Tomlinson
© 1999 Birkhäuser Verlag Basel/Switzerland

Alternative heme degradation pathways

Under normal physiological conditions HO is the major degradation pathway for heme, however alternative heme degradatory pathways do exist in mammalian cells. Xanthine oxidase and H_2O_2 plus nicotinamide adenine dinucleotide phosphate (NADPH) cytochrome P-450 reductase have been demonstrated to degrade heme with the subsequent production of a milieu of pyrrolic complexes [4]. Of possibly more significance is the report of mitochondrial heme degradation activity in the heart which surpasses that of the HO containing microsomal fraction [5]. The authors of this latter report postulated that this mitochondrial pathway could prevent the accumulation of aberrant hemoproteins in the mitochondria which were generated during cellular respiration. In light of the current knowledge of the role of mitochondrial and cytochrome c release in the initiation of apoptosis [6] this heme degradation pathway may be worthy of further investigation. However, it must be noted that in both these cases CO and biliverdin are not products of heme degradation.

HO biochemistry

Although the catabolism of heme by a microsomal enzyme system was first described in the 1960s and termed HO [7, 8] it was not until 1974 that HO was finally characterised as a distinct protein entity [9, 10].

HO is found in the endoplasmic reticulum (ER) and is mainly localised to the smooth ER fraction. A hydrophobic region in the carboxyl terminus of the protein, which is not required for enzymic activity, anchors the protein to the ER [11]. On binding heme at a 1:1 molar ratio HO becomes a transitory hemoprotein [12]. The initial breakage of the heme molecule can occur at any of the four meso-carbon bridges; in this respect the function of HO is to bind the heme molecule in such a way that the cleavage of the tetra pyrrole occurs at the α position, when a type *b* cytochrome is the substrate [13]. This reaction utilizes 3 molecules of oxygen. The terminal oxygen atoms in biliverdin are derived from two separate molecules, with the third molecule being used as the oxygen atom during the conversion of α-carbon atom to CO [14]. The catabolism of heme by HO also requires the concerted activity of microsomal NADPH-cytochrome P-450 reductase [15].

Transfer of electrons from the pyridine nucleotide, NADPH, to the heme-HO complex occurs after its interaction with reductase. These reducing equivalents are required for the conversion and/or maintenance of the central heme iron in its O_2 binding reduced state (Fe^{2+}) and activation of O_2 for α position attack. To date the intermediate products from O_2 binding to the α carbon bridge, to the final products are not known, however, it is though to involve at least 3 intermediate complexes and 3 oxidation steps [16]. Initially, heme is oxidised to α-meso-hydroxyheme

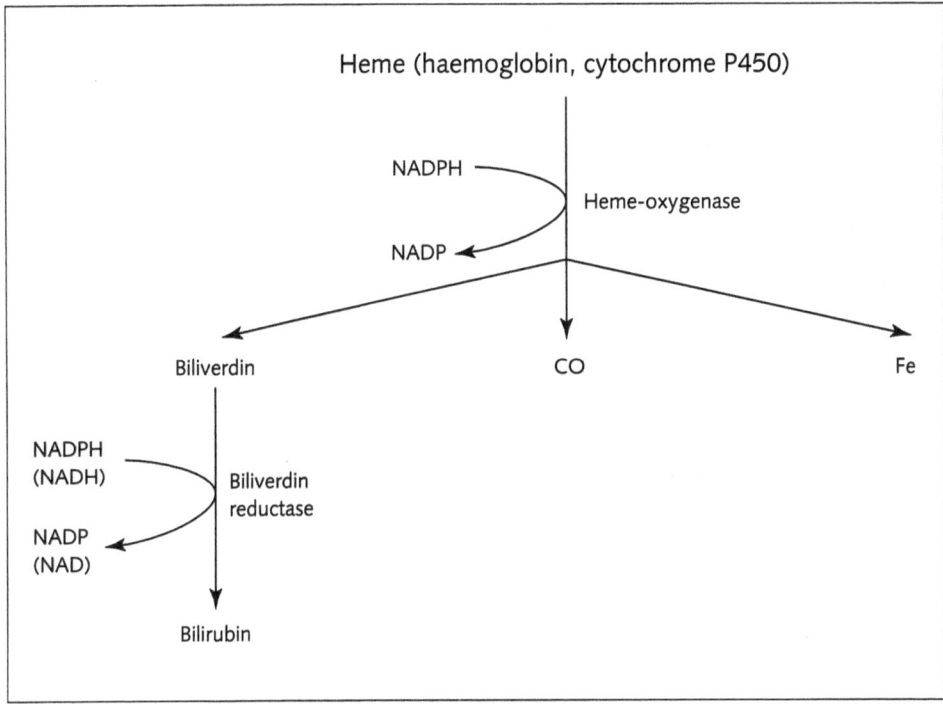

Figure 1
Pathway of heme metabolism. β-Nicotinamide adenine dinucleotide phosphate (NADPH), β-nicotinamide adenine dinucleotide (NADH).

which is itself oxidised to an uncharacterised enzyme bound intermediate, possibly verdoheme, with the subsequent release of CO. Finally, the intermediate is converted to Fe-biliverdin IXα, or biliverdin IXβ under conditions of oxidative stress, from which the iron atom is released [17].

Co-expressed with HO is the enzyme biliverdin reductase. Indeed it has been suggested that a HO-biliverdin reductase-NADPH cytochrome P-450 reductase ternary complex exists *in situ* [18]. This cytosolic enzyme catalyses the two electron reduction of the HO product biliverdin to bilirubin. This enzyme is unique in that it has a dual pyrimidine nucleotide and dual pH dependent points [19]. The enzyme is also heat stable at the protein and mRNA level [20]; with numerous biliverdin reductase isoforms being demonstrated to exist in mammalian tissues [21]. Although the biological significance of having multiple biliverdin reductase isoforms is not known, it appears that they have differing kinetics and specificities for biliverdin IXα and biliverdin IXβ [21]. Indeed, oxidative dimerization of biliverdin reductase, results in a dimeric enzyme which can reduce biliverdin IXβ more efficiently than monomeric biliverdin reductase, thus preventing biliverdin IXβ accumulation [22]. *Figure 1* shows the HO/biliverdin reductase cascade.

Normally, HO derives its substrate from the freshly synthesised pool of heme which has yet to incorporate into relevant apoproteins [23] and from proteins in which the heme moiety is only loosely bound. These include α and β haemoglobin, heme-hemopexin complexes and various cytochrome P-450s [216, 217]. Under pathophysiological conditions the availability of heme greatly increases. ROS and RNS which are generated during oxidative stress, denature various hemoproteins with the subsequent release of heme and disruption of cellular membranes resulting in an influx of extracellular hemoglobin derived heme into the cell [24]. Under these conditions the removal of high intracellular heme concentrations becomes essential, both to allow the reutilisation of the central iron molecule and prevent excessive free radical generation, an intrinsic property of heme molecules.

Substitution of the iron atom of heme with Sn, Zn, Cr or Mn results in a metalloporphyrin which acts as a potent competitive inhibitor of HO [25, 26]. This ability of HO to bind other metalloporphyrins but not utilise them as substrates has been used experimentally to inhibit HO activity *in vitro* and *in vivo* and represents a useful if not specific tool [27, 28].

To date HO has been identified in mammals, birds, frogs, fish, insects, plants and red algae [12]. Recently HO was characterised in *Corynebacterium Diphtheriae* where it is believed to be involved in iron reutilisation [29]. As with other cytochrome P-450-dependent systems, tissue HO concentration fluctuates during animal development, with HO activity being high in fetal organs and decreasing with age [30, 31]. HO activity has also been reported to be higher in female rat adrenal glands than male, which suggests sex differences for the enzyme [32].

Although HO activity has been detected in virtually every mammalian tissue tested to date its activity under physiological conditions varies greatly from organ to organ [12]. Particularly high levels of HO activity are found in the spleen, which probably reflects the high concentration of heme encountered in this organ due to it being the major site of erythrocyte degradation [33]. High levels of HO activity are also found in the brain and testis under normal physiological conditions, however its role in these organs is somewhat more obscure and may relate to a cellular regulatory role of CO [34].

HO isoforms

Initial experiments investigating HO activity took advantage of the high level of HO activity in the spleen and the fact that liver HO activity was dramatically increased after cobalt chloride treatment *in vivo* [9, 10]. The use of these methods and tissues set a precedence for future purification experiments, with the outcome that the 32 kDa inducible isoform of HO, heme oxygenase-1 (HO-1), was the first isoform to be isolated. However, observations that the relatively high HO activity encountered in the rat testis did not increase in response to HO inducing agents lead to the hypothesis that at least two forms of the enzyme existed [35]. Subsequently, a second 36 kDa isoform, heme oxygenase-2 (HO-2), was isolated by using untreated rat brain, testes and liver tissues, and found to be constitutively expressed in many organs under normal physiological conditions [36–40]. Further studies demonstrated that these two isoforms of HO were the products of two different genes [41, 42] and that human tissues also contained both isoform of the enzyme [43, 44].

Under normal physiological conditions HO-2 is the major HO isoform found in mammalian tissues, in an approximate 2:1 ratio to HO-1. Particularly high levels of HO-2 are found in the brain and testis correlating with high HO activity within these organs [38]. In contrast, with the exception of the spleen were HO-1 expression is constitutively high, under physiological conditions HO-1 expression is low.

Although at the biochemical level HO-1 and HO-2 have similar K_M values for heme, 0.24 μm and 0.4 μm respectively, and similar cofactor/coenzyme/substrate profiles, differences do exist between the isoforms at the mRNA and protein level. HO-1 is coded for by a single 1.8 kb transcript, whereas at least 2 transcripts of 1.3 kb and 1.7–1.9 kb have been reported for HO-2 [38, 45, 46]. The occurrence of these two transcripts for HO-2 has been reported to be as a result of differences in polyadenylation with the smaller of the two 1.3 kb, being more efficiently translated than the larger transcript [43].

At the protein level HO-1 and HO-2 share a 43% homology in their primary amino acid sequence. However, a 24 amino acid region within the two isoforms, which forms the hydrophobic heme binding pocket is highly conserved [47]. This portion of the sequence is now recognized as the HO signature (GenBank). Further analysis of the heme binding pocket of HO isoforms by site direct mutagenesis techniques has demonstrated that a histidine residue within the heme pocket of HO-1 (His132) and HO-2 (His151) is essential for enzymic activity [11, 48]. In comparison to the lack of homology between HO-1 and HO-2 isoforms, the evolutionary homology within the same isoform is highly conserved. For instance, rat and human share a greater than 80% amino acid homology for HO-1 and more than 90% homology for HO-2.

Recently a third isoform of HO, termed HO-3 has been reported in the rat [49]. This isoform is a product of a single 2.4 kb transcript and has an apparent molecular weight of 33 kDa. HO-3 was demonstrated to be ubiquitously expressed with high levels of this isoform found in the brain and testes of rats and in this regard is similar to HO-2. Indeed, the amino acid sequence for rat HO-3 and HO-2 share a greater than 90% homology, although antibodies to both HO-1 and HO-2 did not cross react with this third isoform [49]. Intriguingly HO-3 was a poor catalyst for heme degradation and did not have a completely conserved HO signature. These data may indicate an additional role to that of heme catabolism.

Regulation of HO isoforms

Analysis of the upstream promoter region of HO-1 has demonstrated the presence of a number of known transcription factor regulatory sites which include, a TATA box [50]; an upstream stimulatory element (Max/USE) [51]; a metal regulatory element (MRE) [52]; an AP-1 site [53]; an AP-2 and NF-κB site [54]; a cis-regulatory Δ-12 prostaglandin J$_2$ (Δ-12PGJ$_2$) responsive element [55, 56]; an interleukin-6 (IL-6) regulatory element [57]; and a heat shock element (HSE) [50].

Recently 2 hypoxia responsive elements have also been identified in the upstream region of the mouse HO-1 gene [58]. These elements were demonstrated to bind hypoxia-inducible factor-1 (HIF-1). This transcription factor has been shown to increase the expression of various hypoxia-induced genes on exposure of mouse smooth muscle cells to hypoxia and to initiate the expression of HO-1 [59]. In comparison, the HO-2 gene lacks these promoter elements, with only a functional glucocorticoid responsive element being identified upstream of the gene to date [45, 60]. Therefore, analysis of the gene promoter regions of HO isoforms appears to demonstrate, that while HO-2 is a constitutively expressed protein in many cell and tissue types, HO-1 is a highly inducible, highly regulated gene. However, much work is required to determine the relative contribution of these regulatory elements to HO-1 expression during pathophysiological conditions.

HO-1 as a heat shock protein

The cellular response to a variety of stressful stimuli includes the rapid and transient expression of gene products termed 'heat shock' or 'stress' proteins (hsp(s)). In high intracellular concentrations they afford cyto-protection, possibly by protecting protein conformation and prime cells to survive a second stimulus which would otherwise be lethal. Stressful stimuli arise from pathophysiological sources; viral/bacterial infection, inflammation, oxidative stress, ischemia, hypoxia and inflammatory cytokines and from the environment; heat, UV radiation and heavy metals. Stress proteins also perform routine cellular functions under normal physiological conditions [61].

The occurrence of an HSE within a gene's promoter region is indicative of the product of that gene being an hsp. Two HSE, HSE-1 and HSE-2, have now been shown to be present in the upstream promoter region of HO-1 which is consistent with this isoform of HO being an hsp [62]. Furthermore, several studies have shown that HO-1 is identical to hsp32 (32 derived from the molecular weight) expressed after heat and heavy metal treatment of rat hepatoma cells [63], ultra violet (UV) radiation, hydrogen peroxide and sodium arsenite treatment of human skin fibroblasts [64], heavy metal treatment of HeLa and HL60 human cell lines [65] and heavy metal treated, heat stressed and erythrocyte-phagocytosing human macrophages [66, 67]. In addition, the hypoxia-induced 33 kDa oxygen-regulated protein (ORP33) in Chinese hamster ovary cells was also demonstrated to be HO-1 [68].

Although HO-1 is classified as an hsp differences between other classical hsps do exist. Recent work suggests that heat shock transcription factor (HSF) binding to HSE causes only a moderate increase in rat HO-1 expression and does not increase human HO-1 at all [69]. This report also demonstrated that although HSF bound to the HSE within the HO-1 promoter region, a downstream silencer prevented transcription of the gene. In addition, the HO-1 gene does contain introns which appear to be absent in other classical hsps. This lack of introns is hypothesised to contribute to the resistance of hsps to heat shock inhibition of protein synthesis.

HO in inflammatory pathologies

For many years the biological role of HO in cells and tissues was believed to be confined to the rate limiting enzyme responsible for heme catabolism and iron reutilization, with the subsequent release of the waste products biliverdin, free iron and CO. However, it is now becoming increasingly clear that heme catabolism and HO-1 expression may have a more pro-active role to play in pathologies which have inflammatory components.

Acute and chronic inflammation

Many reports have demonstrated that HO-1 can be upregulated in response to inflammatory mediators *in vitro* (see Tab. 1, 2, 3 and 4) and is increased in the liver of mice after IL-1, tumour necrosis factor α (TNFα) and IL-2 treatment [70–72]. To investigate the biological significance of HO expression in acute inflammation we have used a model of acute complement-dependent (alternative pathway) pleural inflammation in the rat.

Injection of carrageenin into the rat pleural cavity results in the development of an acute inflammatory reaction that is maximal at 24 h as assessed by exudate volume and inflammatory cell number, thereafter the inflammatory lesion resolves. The early phase of this reaction is dominated by polymorphonuclear cells (PMNs), primarily neutrophils, followed by an influx of mononuclear cells (MNs) from 12 h onwards. HO activity and exudate bilirubin concentration was found to be highest during the resolution phase of this inflammatory lesion. Increased HO activity correlated with increased HO-1 expression in MNs but not PMNs, which remained negative for both HO-1 and HO-2 throughout the time course. HO-2 expression remained relatively constant throughout the time course. To further investigate the role of the increase in HO-1 expression during inflammatory lesion resolution, we prophylactically administrated either tin protoporphyrin dichloride (SnPP), an HO

Table 1 - Inflammatory cytokines which induce HO-1 mRNA or HO-1 protein expression in vitro.

Cytokine	Cell type: Notes	Reference
IL-1	Rat pancreatic islet cells	[169]
	Rabbit corneal epithelial cells	[170]
IL-1α	Human endothelial cells: HO-1 expression associated with AP-1 activation	[171]
IL-1β	Rat aortic smooth muscle cells: IL-1β induced expression of HO-1 was via a NO independent pathway	[137]
	Rat mesangial cells: HO-1 induction was inhibited by PGE_2	[172]
	Rat pancreatic islet cells	[173, 174]
IL-6	Rabbit corneal epithelial cells	[170]
	Human hepatoma cells (HepG2)	[175, 176]
	Human hepatoma cells (Hep3B)	[57, 177–179]
IL-11	Human hepatoma cells (HepG2)	[176]
TNFα	Rabbit corneal epithelial cells	[170]
	Human endothelial cells: HO-1 expression associated with AP-1 activation	[171]
	Human umbilical vein endothelial cells (HUVEC): HO-1 co-expressed with ICAM-1	[180]
TGFβ	Rabbit corneal epithelial cells	[170]
	Human retinal pigment epithelial cells	[181]
	Bovine choroid fibroblasts	[181]
IFNγ + LPS	Mouse lymphocytic leukaemia cells (L1210): HO-1 mRNA increase simultaneous with that of iNOS mRNA increase	[182]
	Porcine endothelial cells	[183]
	Mouse monocyte/macrophage cells (RAW264.7): iNOS expression preceded HO-1 expression	IHD
LPS	Human hepatoma cells (Hep3B)	[184]
	Rat microglia and astrocytes: Increased HO-1 production was NO dependant	[185]
	Mouse monocyte/macrophages cells (RAW264.7): AP-1 activation implicated in HO-1 up regulation	[132]
	Mouse myeloblastic cells (M1): Oxidative activation of NFκB implicated in HO-1 up regulation	[186]

IHD, in house data.

Table 2 - Reactive oxygen & nitrogen species which induce HO-1 mRNA or HO-1 protein expression in vitro.

Reactive oxygen & nitrogen species	Cell type: Notes	References
SNP, SNAP, SIN-1	Vascular smooth muscle cells	[187]
	Porcine endothelial cells: Ability to induce HO-1 SNP>SNAP>SIN-1. SNAP & SIN-1 protected against H_2O_2 toxicity	[183]
SNP, SIN-1, GSNO	Human cervical cancer cells (HeLa)	[183]
	Human glioblastoma cells (T98G)	[183]
SNP, SNAP	Bovine aortic endothelial cells: Peroxynitrite ($^-$OONO) and super oxide (O_2^-) was found to contribute to NO donors induced HO-1 expression	[188]
SNP	Human gliblastoma cells (A172): No effect of the cytokine mixture IL-1β/TNFα/IFNγ	[189]
	Human epidermal keratinocytes: HO-1 involved in keratinocyte proliferation	[189]
SIN-1	Human endothelial cells: HO-1 production protected cells against TNFα cytotoxicity	[190]
SNAP	Rat microglia cells and astrocytes	[185]
	Sheep pulmonary artery endothelial cells	[191]
SNN	Vascular smooth muscle: HO-1 induction via an cGMP independent mechanism. No effect of peroxynitrite	[192]
GSNO	Rat mesangial cells	[193]
H_2O_2	Rat pancreatic islet cells	[173]
	Rabbit corneal epithelial cells	[170]
	Mouse pancreatic islet cells (beta TC3)	[194]
	Rat acinar cells (AR42J)	[194]
	Human melonma cells (MM96E, MM253cl)	[195]
	Mouse osteoblastic cells	[196]
	Human skin fibroblasts: Not inducible in keratino-cytes which had high constitutive HO-2 expression	[64, 197, 198]
	Human lymphoblastoid cell lines (TK6, L1, L2, L4)	[198]
	Human skin fibroblasts (FEK4): HO-1 induction associated with glutathione depletion	[199]
	Renal epithelial cells (BSC-1)	[124]
	Rat neonatal myocytes	[200, 201]
	Human alveolar macrophage: Lavage from patients with interstitial lung disease. Two patients had spontaneously high HO-1 expression which was associated with phagocytosis	[76]
	Mouse peritoneal macrophages	[202]

IHD, in house data; SNP, sodium nitroprusside; SNAP, S-nitro-N-acetyl penicillamine; SIN-1, 3-morpholinosyalnomine;GSNO, S-nitroso-L-glutathione; SNN, spermine NONOate

Table 3 - Inflammatory mediators which induce HO-1 mRNA or HO-1 protein expression in vitro.

Inflammatory mediators/mechanisms	Cell type: Notes	References
TPA	Human skin fibroblasts	[198]
	Mouse osteoblastic cells	[196]
	Mouse myeloleukemia cells (M1): Increase in HO-1 expression was associated with AP-1 and NFκB activation	[53, 186]
PMA	Rat mesangial cells	[172]
PGA$_1$	Mouse myoblast cells: No induction of hsp70	[203]
PGA$_2$	Human diploid fibroblasts	[204]
Δ^{12}PGJ$_2$	Porcine aortic endothelial cells	[205, 206]
	Rat basophilic leukemia cells (RBL)-2H3	[55]
Erythropagoctosis	Human monocyte/macrophages (U937)	[66]
Midly oxidised LDL	Human aortic endothelial & smooth muscle co-cultures: HO activity prevented monocyte chemotaxis in response to highly oxidised LDL	[102]
Oxidised LDL	Renal tubular epithelial cells (LLC-PK1)	[207]
	Porcine smooth muscle cells	[208]
	Mouse peritoneal macrophages	[209]

IHD, in house data

inhibitor, or ferriprotoporhyrin IX chloride (FePP, hemin), an inducer of HO-1 expression, and investigated the effect on inflammatory parameters at 6 and 24 h. No effect of these drugs on inflammatory parameters was seen at 6 h. However, at 24 h HO inhibition resulted in an 86% increase in exudate volume whereas pre-induction of HO-1 resulted in a 90% and 80% reduction in exudate volume and inflammatory cell number respectively, thus suggesting that HO-1 has an anti-inflammatory effect in this model [73, 74].

We have also investigated the expression of HO isoforms in two immune medi-ated models of acute inflammation [75]. Increased HO activity and HO-1 expres-sion were associated with the resolution phase of both a delayed type hypersensi-tivity (methylated bovine serum albumin-mediated type IV, T cell mediated) model of pleural inflammation, at 48 h, and an immediate type hypersensitivity (bovine serum albumin-mediated type III, Arthus reaction, antibody mediated) pleural model, at 12 h. Whereas immunocytochemical analysis of cell smears from the delayed type reaction showed a similar profile of cell staining for HO-1 to that of

Table 4 - Report on increased HO-1 mRNA or HO-1 protein expression in vitro *after hypoxia or hyperoxia*

Hyper/hypoxia	Cell type: Notes	Reference
Hyperoxia	Human umbilical vein endothelial cells (HUVEC)	[210]
	Epithelial, fibroblasts, macrophages, smooth muscle cells: HO-1 induction associated with AP-1 activation	[211]
	Rat pulmonary micro vascular endothelial cells	[212]
	Chinese hamster fibroblasts (HA1, OC14, O_2R95): Resistance to hyperoxia in O_2R95 cells associated with constitutively high HO-1 expression	[213]
Hypoxia	Rat aortic vascular smooth muscle cells: HO-1 induction mediated by HIF-1	[58]
	Rat aortic and pulmonary smooth muscle cells: Increased CO production increased cGMP levels	[134, 135]
	Rat myocytes	[200, 214]
	Chinese hamster ovary cells	[68]

the carrageenin pleurisy, namely exclusively restricted to MNs, cell smears from the immediate type reaction revealed both MNs and PMNs were positive for HO-1. The reason for the preferential expression of HO-1 in PMNs from an immediate type hypersensitivity reaction is not known. However this observation may relate to the involvement of immunoglobulin G (IgG) and C1q in the antibody (classical) pathway of complement activation which mediates this reaction. Both, IgG and C1q are known to stimulate phagocytosis, a cell function which has been demonstrated to induce HO-1 expression in macrophages [66, 76]. Alternatively, a mediator which inhibits HO-1 expression by PMNs may be present in the carrageenin and delayed type hypersensitivity-mediated acute inflammatory pleurisies but absent in the immediate type inflammatory lesion. Taken together these results demonstrated that HO and in particular HO-1 may have a role to play in the resolution of an acute inflammatory reaction.

In addition to our work, other groups have demonstrated a role for HO in ocular, kidney and lung inflammation. Kappas and colleagues [77] have used a 6 day contact lens-induced corneal inflammatory model in the rabbit to investigate the role of HO in inflammation. These workers demonstrated that impregnation of contact lenses with SnCl2 resulted in a marked increase in HO-1 mRNA and HO activity in corneal epithelium in comparison to vehicle control treated lenses. The increase in HO-1 with SnCl2 was associated with a 18% decrease in corneal thickness and a 73% decrease in a subjective (corneal edema) inflammatory score.

In the kidney, FePP has recently been demonstrated to suppress inflammatory cell influx in heterologous nephrotoxic nephritis and accelerated nephrotoxic nephritis, with proteinurea and glomerular thrombi being significantly decreased by HO inducer treatment [78]. Nath and colleagues [79] have also demonstrated that nephrotoxic serum-induced glomerular inflammation results in an marked increase in kidney HO-1 mRNA and positive staining of renal tubules for HO-1, 6 and 24 h after administration. The induction of HO-1 appeared not to be related to increased availability of free heme as there was no evidence of increased hematuria or increased haemoglobin/heme excretory rates. The 24-hour prior administration of nephrotoxic serum also significantly decreased glycerol-mediated kidney damage at 48 h, improving both functional (creatinine excretion) and morphological parameters of kidney damage. As the protection afforded by prior administration of nephrotoxic serum against glycerol-mediated injury was attenuated by administration of the HO inhibitor SnPP, the authors concluded that increased HO-1 expression induced by glomerular inflammation protects renal tubules against tissue injury. In a previous study these authors had shown that pretreatment with low doses of endotoxin also improved kidney function and antioxidant status in response to glycerol-induced kidney damage [80]. This protection was also associated with increased HO-1 and ferritin expression and could be inhibited by the administration of HO inhibitors.

However, the protective effect of nephrotoxic serum did not extend to an ischemic reperfusion kidney injury model. To understand this apparent dichotomy one must examine the expression of HO in these two models. Glycerol-mediated renal damage is a model of rhabdomyolysis, a disease mediated by excessive myoglobin and hemoglobin release. HO, as well as ferritin was upregulated in this model. Inhibition of HO exacerbated renal disfunction, whereas induction of HO by hemoglobin prior to the onset of myolysis and hemolysis, protected kidney function and reduced mortality [81]. In contrast HO is not induced after ischemic reperfusion in the kidney and prior induction of HO expression does not protect the kidney from ischemic reperfusion injury [79, 82, 83]. It is therefore not surprising that HO modulation has no effect in this model as HO appeared not to be induced.

HO is also induced during cisplatin-induced nephrotoxicity, a stimulus which is not associated with increased cellular levels of free heme; again inhibition of HO activity exacerbated kidney damage [82].

Finally, two papers have demonstrated induced HO-1 expression in ozone- and virus-induced lung inflammation [84, 85]. In the viral-induced lung inflammation, intra-nasal administration of influenza virus A/PR/8/34 (H1N1) increased HO-1 expression and HO activity to a greater extent in the lungs of C57Bl/6 mice, which are resistant to viral-induced mortality, in comparison to the lungs of C_3H/HeJ mice, which are susceptible to virus-induced death. This observation is in line with previous reports demonstrating differences in HO-1 expression between mouse strains in response to an atherogenic diet [86, 87].

To date few studies have addressed the role of HO in chronic inflammatory tissue. We have used the murine chronic granulomatous tissue air pouch model to determine the expression of HO isoforms. HO-2 expression was found to occur through out the time course (24 hours to 28 days) of this model. In comparison, the highest level of HO-1 mRNA, protein and HO activity was found at day 3 and was coincident with the expression of HO-1 by influxing MNs. Smooth muscle cells were also found to be positive for HO-1 throughout the time course of this model. The expression of HO-1 in granulomatous tissue at 3 days is in line with HO having a role in the transition of acute inflammation to chronic inflammation. Daily administration of the HO inhibitor had no effect on the maximal weight of the granuloma at 7 days, but significantly reduced an index of angiogenesis (see chapter by Seed et al., this book).

One other study showed hepatic HO activity upregulated during adjuvant-induced arthritis and it was suggested, in combination with decreased heme production, to contribute to the low hepatic cytochrome P-450 levels seen in arthritic animals [88]. Clearly the role of HO in chronic inflammation requires further research.

HO-1 knockout mice and inflammation

Recently two papers on the development and characterisation of heterozygous and homozygous HO-1 knockout mice have been published, with initial data supporting a cyto-protective/ anti-inflammatory role for HO-1 [89, 90]. The major finding of these studies was that HO-1 knockout mice accumulated iron in hepatic and renal tissue but decreased serum iron levels. The authors speculated that at least two different heme catabolic pathways exist. The first pathway which is HO-1-dependent contributes to the extracellular release of iron, while a second pathway results in the intracellular storage of iron liberated from heme. This second pathway of heme catabolism has to date not been characterised, however initial data would exclude an HO-2-mediated pathway [91]. Although alternative heme degradatory pathways do exist (see above), it will be interesting to see to what extent, if any, the newly described HO-3 has a role in heme catabolism.

In addition to the accumulation of iron it was shown that from around 20 weeks of age HO-1 deficient animals have a increased ratio of CD4+:CD8+ T lymphocytes (with many activated CD4+ present). They displayed increased splenic weight and had significantly higher levels of oxidized protein and lipid peroxidation in liver and kidney compared to wild type controls. HO-1 knockout mice also developed a progressive chronic inflammatory disease characterised by inflammatory cell infiltration of the liver with an increase in the adherence of monocytes to the vasculature [89]. Interestingly this pathology is similar to the anaemia of chronic diseases, a phenomenon which often occurs in patients suffering from chronic infections, malignant tumours or autoimmune diseases. The use of young, gross pathology- deficient, HO-1 knockout mice should help elucidate the significance of HO-1 in both acute and chronic inflammation.

HO-1 as an anti-inflammatory agent

The mechanism by which HO-1 elicits its cyto-protective/ anti-inflammatory action is not known, however a primary mode of action must be the catabolism of free heme which is a potent generator of ROS [92]. In addition the products of this catabolism may also contribute to the cyto-protective/ anti-inflammatory action of HO-1. Physiological concentrations of free and albumin-bound bilirubin have been shown to prevent the oxidation of albumin and lipoic acid bound to albumin, oxidation of liposomes [93–96] and the oxidation of low density lipoprotein (LDL) [218] by peroxyl radicals. At low partial pressures of oxygen, bilirubin was found to be a better inhibitor of lipid peroxidation in liposomes than vitamin E [97]. Albumin-bound bilirubin has also been shown to prolong the survival of human myocytes after xanthine oxidase treatment [97]. One of the resulting products of bilirubin oxidation by two molecules of peroxyl radicals, is biliverdin [96]; which also has antioxidant properties [95, 96, 98, 99]. The bile pigments may also protect proteins from free radical attack. Recently it has been shown that bilirubin, as well as other endogenous anti-oxidants, significantly improved the anti-inflammatory effect of low dose dexamethasone in a model of ischemic/reperfusion-mediated inflammation in the mouse paw [100]. The mechanism by which bilirubin enhances dexamethosone anti-inflammatory activity is not known, however it has been reported that the glucocorticoid receptor is extremely sensitive to free radical inactivation [101]. Whether the bile pigments can reverse this inactivation is unknown, nevertheless the ability of HO activity to modulate steroid activity at the cellular level represents an intriguing possibility.

It is also noteworthy that in addition to the steroids other anti-inflammatory drugs may interact with the HO cascade. Work carried out in our department has demonstrated that various non-steroidal anti-inflammatory drugs, including aspirin, potentiate HO-1 expression in the rat carrageenin pleurisy model at relevant doses. In addition, the anti-rheumatic disease-modifying drug gold thiomalate, has the ability to inhibit inducible nitric oxide synthase (iNOS) expression in a mouse monocyte/macrophage cell line (RAW246.7) which is associated with increased HO-1 expression (in house data).

Ishikawa and colleges have used a human endothelial and smooth muscle cell co-cultures to demonstrate that the HO-1 induction in endothelial cell in after treatment with mildly oxidised LDL was associated with an inhibition of monocyte transmigration. Furthermore inhibition of monocyte transmigration was also accomplished by pretreatment of the co-cultures with biliverdin or bilirubin, whereas pretreatment of the co-cultures with the HO inhibitor SnPP augmented monocyte transmigration in the same co-culture system. It was concluded that HO-1 induction may represent a protective mechanism, via the increased production of the antioxidant bile pigments, during atherosclerosis or restenosis [102]. Indeed, high and low serum bilirubin levels are associated with decreased and increased coronary artery disease respectively and increased HO-1 production has been reported to occur during atherosclerosis [219–221]. In addition to the antioxidant properties of bile pigments, both biliverdin and conjugated bilirubin have been demonstrated to have anti-complement effects *in vitro* and significantly prolonged the survival time of guinea pigs in which Forssman's anaphylaxis has been induced [103]. Bilirubin has also been reported to inhibit the cytotoxic activity and IL-2 production from human T lymphocytes [104–106]. In this regard it is interesting to note that in our inflammatory models the highest exudate bilirubin concentration was recorded during the resolution phase in an antibody-mediated acute inflammation. However, further work will be required to determine the relationship between the complement cascade and the bile pigments.

The cyto-protective/anti-inflammatory effects of HO-1 expression may also be due to the modulation of other inflammatory enzyme cascades. CO can bind and inhibit heme-containing enzymes which are involved in inflammatory reactions. Alternatively, high HO activity may deplete the intracellular heme pool thus inhibiting the activity of heme-dependent enzymes by preventing heme apoprotein complexes forming; two such targets include the COX and NOS enzyme systems. Indeed, increased HO activity is associated with a decrease in cytochrome P-450-dependent arachidonic acid metabolites in FePP-treated hypertensive rats and a model of corneal inflammation [107, 108].

NO donors have been shown to induce HO-1 expression (see Table 2) while inhibiting HO activity [109]. With the use of various inflammatory models *in vivo* we have also found a clear temporal relation between NOS and HO, with iNOS expression always proceeding that of HO-1. Again, these data provide initial evidence for an interaction between two inflammatory enzyme cascades, namely NOS and HO, during inflammation (see chapter by Tomlinson and Willoughby, this book).

While CO has also been shown to have effects similar to that of NO including prevention of platelet aggregation and increasing cGMP levels via guanylate cyclase [110] a recent report has also demonstrated that increases in cGMP in cultured neurons after NO treatment can be inhibited by physiological concentrations of CO [111]. Recently, we have demonstrated that the inhibition of increased expression and activity of iNOS after treatment with peroxisome proliferator-activated receptor γ (PPARγ) agonists of the prostaglandin J series in activated mouse macrophages, is partially due to the increased levels of HO-1 [222]. PPARγ agonists have been demonstrated to inhibit the expression of TNFα, IL-1β and IL-6 and the activation of AP-1, STAT and NFκB [112–114]. Whether HO-1 expression can also account for some of these other reported activities is currently under investigation. The hypothesis that HO-1 expression may inhibit inflammatory cytokine release is supported by data demonstrating the inhibition of TNFα production from the human monocytic cell line THP-1, by FePP and this can be prevented by the HO inhibitor SnPP [115].

It is obvious that from the initial data obtained interactions between HO, COX and NOS enzyme systems do exist. Although further work is required, these data demonstrate that the products of two enzymes involved in the developmental stage of acute inflammatory lesions, iNOS and COX, can lead to an increase in HO-1 which is associated with the resolution of acute inflammation and cyto-protection. Therefore, it is feasible that whilst inhibition of COX isoforms and iNOS will have beneficial effects in the short term for acute inflammatory lesions, inhibition of COX and iNOS may prevent the production of mediators which stimulate endogenous anti-inflammatory cascades and therefore ultimately prevent the resolution of acute inflammation.

It would be expected that increased intracellular iron, generated by heme catabolism by HO, would increase the generation of ROS via the Haber-Weiss reaction. However, increased catabolism of heme by HO is also associated with an increased production of ferritin which would then chelate the free heme [80, 81, 116]. Ferritin has also been reported to have anti-inflammatory activity [117]. Indeed, it has been suggested that HO-1 induction is not the predominant cyto-protective agent, but represents an important intermediate step in the increase of ferritin production which is the true cyto-protective agent [116, 118, 119]. However several papers have reported cyto-protective effects of HO-1 expression which appear to be independent of ferritin induction [120–122].

Finally it must be noted that HO inhibition is also reported to be cytoprotective. Inhibition of the basal HO activity in the human breast adenocarcinoma cell line (MCF-7) by SnPP, attenuated the DNA strand breaks and growth inhibition encountered after menadione treatment [123]. The HO inhibitors zinc 2, 4-bis-glycol protoporphyrin was also shown to inhibit the reduction of cell viability in renal epithelial cells (BSC-1) seen after treatment with hemin for 1 h, although the authors of this paper did see an exacerbation of the cytoxic effects of 24 h treatment by hemin when the HO inhibitor was added to cultures [124]. The significance of these results is not clear, however it interesting to note that the beneficial effects of HO inhibition on cellular function after a stressful insult did occur when the predominate isoform present was likely to be HO-2 and not HO-1. Taken together these results suggest a role for HO-2 in the cells response to oxidative stress.

HO-2 in inflammation

Although the HO-2 knockout mouse has been characterised has having no major physiological problems [91] it has recently been demonstrated that homozygous

HO-2 knockout mice are more susceptible to hyperoxia-induced oxidative stress injury in the lung [125]. This report demonstrated that while HO-2 knockout mice had a higher basal level of HO-1 and higher HO activity in the lung before and after hyperoxia, their ability to further induce lung HO-1 expression and ferritin in response to hyperoxia was impeded. The authors concluded that HO-2 functions to augment the turnover iron during oxidative stress with HO-1 unable to compensate for the lack of HO-2, and clearly demonstrated that HO-2 may have a role to play in other pathophysiological conditions.

The physiological role for HO-2 is believed to be in heme homeostasis. However, Maines has recently suggested a role for HO-2 in cellular heme/oxygen sensing [34]. The hypothesis is based on the occurrence of an oxygen-sensing motif in the 3' untranslated region of the HO-2 mRNA and the presence of two high affinity heme binding regions termed heme regulatory motifs (HRMs) in addition to the catalytic heme pocket within HO-2 protein. The affinity of heme for these HRMs far exceeds that of the catalytic pocket, but has no intrinsic catalytic activity [126]. At this point it must be noted that HO-1 does not contain HRMs whereas the recently described HO-3 contains 2 HRMs [49]. It is hypothesised that in response to oxidative stress the availability of free heme within a cell increases, this leads to the binding of heme to the HRMs of HO-2. Heme molecules have an intrinsic ability to generate ROS, however binding of heme to the HRM may facilitate heme-mediated ROS production by preventing degradation by HO catalytic activity. Although in the short term increased ROS production would lead to the increased expression of pro-inflammatory mediators, increased ROS generation by the HO-2 HRM-heme complex would also lead to an increased HO-1 expression, probably via NFκB an AP-1 activation. Finally, as the intracellular heme content of the cell was reduced by the increase in HO activity, the generation of ROS by the HO-2 HRM-heme complex would eventually stop due to the lack of heme. This mechanism would allow HO-2/HO-1 to tightly control intracellular heme concentration, whereas inhibition of the heme HO-2 HRM interaction would prevent the induction of HO-1, resulting in intracellular heme accumulation.

It is also possible that the ROS generating ability of HO-2 HRM-heme interactions may have a wider implication for cell signalling during inflammation. Excessive release of ROS and RNS has been demonstrated to be responsible for the tissue damage seen in inflammation. However of possibly greater significance is the ability of ROS and RNS to meditate transcription factor activation [127, 128] Increased cytosolic ROS and RNS concentrations can activate a number of transcription factors including NFκB and AP-1 [129]. In comparison, high concentration of ROS and RNS or antioxidant depletion within the cell nucleus inhibits the binding of transcription factors to their corresponding DNA regulator elements [130]. Therefore, intracellular mechanisms which may regulate ROS production could have profound effects on transcription factor activation in this respect. It is intriguing to note that dexamethasone treatment of HeLa cells increase perinuclear staining for HO-2

[60]. However a general heme scavenging role for the HRMs of HO-2 at this point cannot be discounted.

Endotoxic shock

Endotoxic shock resulting from the release of lipopolysaccharide (LPS) from gram-negative bacteria causes profound hemodynamic changes [131] and major tissue damage as a result of the generation of various ROS and pro-inflammatory media-tors. Due to the reported vasodilatory effects of CO and the possible tissue protective effects of HO-1, the effect of LPS on HO expression and its role in endotoxic shock has drawn some attention. *In vitro* LPS increases HO activity and HO-1 expression in both vascular smooth muscle cells and monocytic cells, probably via NFκB activation and/or AP-1 [132–135]. In addition, LPS administration *in vivo* results in increased hepatic HO activity and HO-1 expression [71, 72]. In an LPS-induced endotoxic shock model of lung damage and sepsis, Choi and colleagues demonstrated a dramatic increase in HO activity and HO-1 mRNA levels in the lungs of LPS-treated animals [136]. Prior to administration of hemoglobin, which induced HO-1 expression, completely protected animals against lethal endotoxic shock, whereas inhibition of HO with SnPP increased the susceptibility of rats to lethal doses of LPS. Hemoglobin pretreatment, but not desferrioxamine, was found to maintain mean arterial blood pressure in response to LPS, prevent hepatic and renal dysfunction and decrease LPS-induced neutrophil influx into the lung [122]. The authors of this latter study concluded that HO-1 induction by hemoglobin pre-vented both the hemodynamic and inflammatory effects of LPS administration and was independent of ferritin production and iron chelation.

However Yet et al., have recently reported that increased HO-1 protein and HO activity in aortic tissue may contribute to the hypotension seen in LPS-treated rats [137]. This study demonstrated that the HO inhibitor ZnPP prevented the hypoten-sion produced by LPS. It is also in line with other reports which have demonstrated that HO inhibition can increase mean arterial pressure whereas HO substrates decrease blood pressure in spontaneously hypertensive rats [108, 138, 139]. How-ever, it is important to note that HO inhibitors and HO substrates appear to have little effect on normotensive animals [137, 138]. This result suggests that HO may only play a role in the control of blood pressure during conditions of compromised pressor responses. Indeed hemodynamic stress is known to upregulate HO-1 expres-sion [140].

To further elucidate the role of HO-1 in endotoxic shock we have assessed the temporal and spatial expression in a rat model in which low doses of LPS are con-stantly infused into conscious animals. This model of endotoxic shock is preferable to models in which bolus doses of LPS are administered to unconscious animals and thus already have altered pressor responses [141]. Although a 6 h infusion of LPS resulted in the induction of iNOS expression in a number of organs including the lung, no effect was seen on HO-1 expression at this time point in comparison to saline-treated rats. However, a 24 h infusion of LPS resulted in increased HO-1 expression in rat lung, liver, kidney, heart, aorta and mesentery. This time point also corresponded to the time of maximal vasodilation in this model. Immunohisto-chemistry revealed that the increase in HO-1 protein in lung, liver, kidney and heart at 24 h was probably related to an influx of HO-1-expressing inflammatory cell in these organs, whereas increased expression in the aorta was related to HO-1 posi-tive vascular smooth muscle [142]. We speculate that the rapid induction of HO-1 serves as an adaptive mechanism to protect cells from oxidative damage encoun-tered in endotoxic shock. However, excessive production of CO by HO isoforms may also contribute to the delayed vasodilatation observed in this model.

Although the pathophysiological role of HO in the vasculature requires further work, several possible mechanisms by which HO could modulate blood pressure can be envisaged. CO has stimulatory effects on guanylate cyclase [110] and has been demonstrated to inhibit the production of the vasoactive agent endothelin-1 [135]. Increased HO activity has been reported to decrease the levels of cytochrome P-450 oxygenase which produce important blood pressure regulating factors [143]. Biliverdin/bilirubin and free iron may modulate blood pressure by regulating the production of ROS [144] and cross talk between the NOS and HO enzyme systems may have an impact on blood pressure control.

Finally, recent work shows that LPS-challenged HO-1 knockout mice had a sig-nificantly higher mortality rate (83%) when compared to heterozygous (6%) and wildtype (0%) litter mates [90] Histological analysis of liver tissue from homozygous HO-1 knockout mice challenged with LPS revealed iron-overloading which was hypothesised to be the cause of mortality. However it must be noted that LPS chal-lenge in normal animals results in a large increase in hepatic HO-1 expression, there-fore the extensive iron-overloading in the HO-1 knockout mouse seen after LPS-treatment may be an artefact of this transgenic model and not of clinical significance.

Transplantation

The rejection of either an allograft or xenograft by the host immune system leads to a pronounced inflammatory response, with the release of many mediators which have been implicated in HO-1 induction. Several groups have now investigated the expression and role of HO-1 in transplantation models. Nath and colleagues demonstrated an increase in HO-1 mRNA and HO activity in allograft kidneys 5 days after transplantation, while control isografts showed no HO-1 induction [145]. Immunohistochemical analysis of kidney sections showed that the increase in HO was due to the accumulation of macrophages in the rejecting kidney which expressed HO-1.

In an attempt to identify protective genes in cardiac xenografts (hamster donor, rat recipient) which have been treated with cobra venom factor and cyclosporin A, Bach and colleagues have compared the expression of various immune, inflammatory and apoptosis-associated proteins in rejecting cardiac xenografts (5-49 day rejection) and non-rejecting or "accommodated" cardiac xenografts (surviving greater than 57 days) [146]. The accommodated xenografts expressed IL-4, IL-10 and IL-13, a cytokine profile consistent with a Th2 like response. They had high IgG2c titres and increased expression (greater than 50%) of A20, bcl-2, bcl-x_L and HO-1 in endothelial and smooth muscle cells. HO-1 expression was also dramatically increased in white blood cells and myocytes in accommodated hearts. In comparison, rejected xenografts expressed IL-2, IFNγ and TNFα, had an increased incidence of apoptosis, and increased expression of Bax, Bad and CPP-32. In rejected heart xenografts, HO-1 expression was restricted to less than 1% of the total endothelial and white blood cells. It was concluded that HO-1 may represent a important protective response in xenograft.

In an attempt to further elucidate the role of HO in transplantation we have used a mouse skin allograft model (C_3H donor to CBA recipients). Initial work has shown that in animals which received vehicle control, transplanted skin normally rejects within 10-11 days, however in animals which were daily administered an HO inducer, FePP, 75% of skin transplants were still viable 22 days after transplantation, but rejected within 2-3 days once FePP administration was withdrawn. Interestingly, leucocyte infiltration of the donor skin in FePP treated animals is similar to that in vehicle-treated animals immediately prior to rejection. It has recently been reported that an immunosuppressive peptide D2702.75-84 which corresponds to the amino acid residues 75 to 84 of the $α_1$-helix of HLA-B2702 and induces long term graft survival in a number of allograft models [147–150], induces HO-1 expression and HO activity *in vivo* [151]. These results appear to indicate that increased HO-1 expression prevents the rejection of skin allografts most probably at a stage after T cell activation. In addition HO-1 expression may also contribute to the revascularisation of donor tissue (see chapter by Seed et al., this book).

Ischemia/reperfusion injury

Ischemic-reperfusion is a complex pathological process which is initiated by the reintroduction of oxygen (reperfusion) into an anoxic/hypoxic (ischemic) tissue. Although several factors may be responsible for initiating and contributing to the response, it is now becoming increasingly apparent that post-ischemic inflammation is a major cause of the tissue damage seen in these pathologies [223].

Although Nath and colleagues did not show a increase in HO-1 expression in response to an ischemic/reperfusion injury, (see acute and chronic inflammation section in this chapter) [83], several other investigators have. HO-1 expression is increased in liver [224, 225], kidney [24, 152], heart [153–156] and brain [157–162].

The possible significance of HO-1 expression in ischemic reperfusion injury is not known. *In vitro*, treatment of cardiac myocytes increased the expression of HO-1 and hsp70 [163] and was associated with the protection of cells against H_2O_2 cytotoxicity and the preservation of glutathione levels. As the generation of ROS and the activation of complement [164, 165] are believed to contribute to ischemic/reperfusion tissue damage one may expect that the anti-oxidant and anti-complement activities of the bile pigments would afford a protective effect. Indeed, Gunn rats have congenital hyperbilirubinemia and are comparatively resistant to hypoxia-induced oxidative injury in comparison to normal animals [226]. Maulik and colleagues have suggested that NO-mediated myocardial preservation in an isolated rat heart model of acute ischemic and reperfusion injury, is partly mediated by a CO-cGMP mechanism [153, 154]. It has also been suggested that CO may relieve the hypoxia encountered during ischemia by relaxing coronary and aortic smooth muscles and dilating coronary arteries [140].

Conclusion

With the continuing realisation of the biological significance of CO, the bile pigments and free iron, research into the role of HO isoforms in pathological conditions has recently gained pace. It is hoped that the arrival of HO-1 and HO-2 knockout mice will greatly facilitate future research. However perhaps of greater benefit to the field would be the development of specific and selective HO inhibitors. Such inhibitors would overcome some of the problems encountered with metalloporphyrin based inhibitors and allow the relative effects of HO isoforms in various conditions to be addressed. Furthermore, investigations into the expression of HO isoforms in clinical samples would help determine the relative importance of this enzyme system in human pathologies

In conclusion, although the catabolism of heme by HO was documented in the late 1960s it is only recently that we have began to understand its biological role in physiological and pathophysiological conditions. The overall assessment of HO to date in inflammatory pathologies is that it represents an endogenous protective mechanism which attenuates tissue damage by preventing the accumulation of deleterious high heme concentrations in tissues and allowing the re-utilisation of iron. Although the role in granuloma formation and angiogenesis will require careful consideration. Induction of HO-1 may also represent an interesting therapeutic strategy in particular in acute inflammatory pathologies and transplantation and ischemic/reperfusion injury. Correspondingly, it is possible that high expression of HO-1 reported in some forms of cancer [166–168] may contribute to tumorogenesis by protecting tumor cells against oxidative stress. Therefore the inhibition of HO activity in cancer may not only increase the susceptibility of tumor cells to chemotherapeutic agents but increase inflammation and decrease angiogenesis.

Finally, it is becoming apparent that besides iron the other two products of heme catabolism by HO, CO and the bile pigments, may also represent important biological mediators and not only waste products.

References

1   Willoughby DA, A Tomlinson, D Gilroy, D Willis (1996) Inducible enzymes with special reference to COX-2 in the inflammatory response. In: J Vane, J Botting, R Botting (eds) : *Improved non-steroid anti-inflammatory drugs: COX-2 enzyme inhibitors*. Kluwer Academic Publishers & William Harvey Press, Dordrecht, Boston, London, 67–83

2   de Haro C, R Mendez, J Santoyo (1996) The eIF-2alpha kinases and the control of protein synthesis. *FASEB J* 10: (12) 1378–1387

3   May BK, SC Dogra, TJ Sadlon, CR Bhasker, TC Cox, SS Bottomley (1995) Molecular regulation of heme biosynthesis in higher vertebrates. *Prog Nucleic Acid Res Mol Biol* 51: 1–51

4   Guengerich FP (1978) Destruction of heme and hemoproteins mediated by liver microsomal reduced nicotinamide adenine dinucleotide phosphate-cytochrome P-450 reductase. *Biochemistry* 17: (17) 3633–3639

5   Kutty RK and MD Maines (1987) Characterization of an NADH-dependent haem-degrading system in ox heart mitochondria. *Biochem J* 246: (2) 467–474

6   Kroemer G, N Zamzami, SA Susin (1997) Mitochondrial control of apoptosis. *Immunol Today* 18: (1) 44–51

7   Tenhunen R, HS Marver, R Schmid (1969) Microsomal heme oxygenase. Characterization of the enzyme. *J Biol Chem* 244: (23) 6388–6394

8   Tenhunen R, HS Marver, R Schmid (1968) The enzymatic conversion of heme to bilirubin by microsomal heme oxygenase. *Proc Natl Acad Sci USA* 61: (2) 748–755

9   Maines MD, A Kappas (1974) Cobalt induction of hepatic heme oxygenase; with evidence that cytochrome P-450 is not essential for this enzyme activity. *Proc Natl Acad Sci USA* 71: (11) 4293–4297

10 Yoshida T, S Takahashi, G Kikuchi (1974) Partial purification and reconstitution of the heme oxygenase system from pig spleen microsomes. *J Biochem (Tokyo)* 75: (5) 1187–1191

11 McCoubrey WK Jr, MD Maines (1993) Domains of rat heme oxygenase-2: the amino terminus and histidine 151 are required for heme oxidation. *Arch Biochem Biophys* 302: (2) 402–408

12 Maines MD (1988) Heme oxygenase: function, multiplicity, regulatory mechanisms, and clinical applications. *FASEB J* 2: (10) 2557–2568

13 Kutty RK, MD Maines (1982) Oxidation of heme c derivatives by purified heme oxy-

genase. Evidence for the presence of one molecular species of heme oxygenase in the rat liver. *J Biol Chem* 257: (17) 9944–9952

14 Docherty JC, BA Schacter, GD Firneisz, SB Brown (1984) Mechanism of action of heme oxygenase. A study of heme degradation to bile pigment by 18O labeling. *J Biol Chem* 259: (21) 13066–13069

15 Trakshel GM, RK Kutty, MD Maines (1986) Cadmium-mediated inhibition of testicular heme oxygenase activity: the role of NADPH-cytochrome c (P-450) reductase. *Arch Biochem Biophys* 251: (1) 175–187

16 Yoshida T, M Noguchi (1984) Features of intermediary steps around the 688-nm substance in the heme oxygenase reaction. *J Biochem (Tokyo)* 96: (2) 563–570

17 Sano S, T Sano, I Morishima, Y Shiro, Y Maeda (1986) On the mechanism of the chemical and enzymic oxygenations of alpha-oxyprotohemin IX to Febiliverdin IX alpha. *Proc Natl Acad Sci USA* 83: (3) 531–535

18 Yoshinaga T, S Sassa, A Kappas (1982) The occurrence of molecular interactions among NADPH-cytochrome c reductase, heme oxygenase, and biliverdin reductase in heme degradation. *J Biol Chem* 257: (13) 7786–7793

19 Kutty RK, MD Maines (1981) Purification and characterization of biliverdin reductase from rat liver. *J Biol Chem* 256: (8) 3956–3962

20 Ewing JF, CM Weber, MD Maines (1993) Biliverdin reductase is heat resistant and coexpressed with constitutive and heat shock forms of heme oxygenase in brain. *J Neurochem* 61: (3) 1015–1023

21 Frydman RB, ML Tomaro, J Rosenfeld, J Awruch, L Sambrotta, A Valasinas, B Frydman (1987) Biliverdin reductase: substrate specificity and kinetics. *Biochim Biophys Acta* 916: (3) 500–511

22 Tomaro ML, J Frydman, RB Frydman (1990) The *in vivo* and *in vitro* oxidation of molecular form 1 of biliverdin reductase to molecular form 3 by diamide. *FEBS Lett* 263: (1) 38–42

23 Maines MD, A Kappas (1976) The induction of heme oxidation in various tissues by trace metals: evidence for the catabolism of endogenous heme by hepatic heme oxygenase. *Ann Clin Res* 8 Suppl 17: 39–46

24 Maines MD, RD Mayer, JF Ewing, WK McCoubrey Jr (1993) Induction of kidney heme oxygenase-1 (HSP32) mRNA and protein by ischemia/reperfusion: possible role of heme as both promotor of tissue damage and regulator of HSP32. *J Pharmacol Exp Ther* 264: (1) 457–462

25 Kappas A, GS Drummond (1984) Control of heme and cytochrome P-450 metabolism by inorganic metals, organometals and synthetic metalloporphyrins. *Environ Health Perspect* 57: 301–306

26 Maines MD (1981) Zinc-protoporphyrin is a selective inhibitor of heme oxygenase activity in the neonatal rat. *Biochim Biophys Acta* 673: (3) 339–350

27 Kappas A, GS Drummond (1986) Control of heme metabolism with synthetic metalloporphyrins. *J Clin Invest* 77: (2) 335–339

28 Vreman HJ, BC Ekstrand, DK Stevenson (1993) Selection of metalloporphyrin heme oxygenase inhibitors based on potency and photoreactivity. *Pediatr Res* 33: (2) 195–200

29 Wilks A, MP Schmitt (1998) Expression and characterization of a heme oxygenase (Hmu O) from Corynebacterium diphtheriae – Iron acquisition requires oxidative cleavage of the heme macrocycle. *J Biol Chem* 273: (2) 837–841

30 Abraham NG, RD Levere, ML Freedman (1985) Effect of age on rat liver heme and drug metabolism. *Exp Gerontol* 20: (5) 277–284

31 Maines MD, A Kappas (1975) Study of the developmental pattern of heme catabolism in liver and the effects of cobalt on cytochrome P-450 and the rate of heme oxidation during the neonatal period. *J Exp Med* 141: (6) 1400–1410

32 Veltman JC, MD Maines (1985) Sex difference in adrenal heme and cytochrome P-450 metabolism: evidence for the repressive regulatory role of testosterone. *J Pharmacol Exp Ther* 235: (1) 71–75

33 Spencer RP, HA Pearson (1975) The spleen as a hematological organ. *Semin Nucl Med* 5: (1) 95–102

34 Maines MD (1997) The heme oxygenase system: a regulator of second messenger gases. *Annu Rev Pharmacol Toxicol* 37: 517–554

35 Maines MD, AS Chung, RK Kutty (1982) The inhibition of testicular heme oxygenase activity by cadmium: A novel cellular response. *J Biol Chem* 257: (23) 14116–14121

36 Braggins PE, GM Trakshel, RK Kutty, MD Maines (1986) Characterization of two heme oxygenase isoforms in rat spleen: comparison with the hematin-induced and constitutive isoforms of the liver. *Biochem Biophys Res Commun* 141: (2) 528–533

37 Maines MD, GM Trakshel, RK Kutty (1986) Characterization of two constitutive forms of rat liver microsomal heme oxygenase. Only one molecular species of the enzyme is inducible. *J Biol Chem* 261: (1) 411–419

38 Trakshel GM, MD Maines (1989) Multiplicity of heme oxygenase isozymes HO-1 and HO-2 are different molecular species in rat and rabbit. *J Biol Chem* 264: (2) 1323–1328

39 Trakshel GM, RK Kutty, MD Maines (1988) Resolution of the rat brain heme oxygenase activity: absence of a detectable amount of the inducible form (HO-1). *Arch Biochem Biophys* 260: (2) 732–739

40 Trakshel GM, RK Kutty, MD Maines (1986) Purification and characterization of the major constitutive form of testicular heme oxygenase. The noninducible isoform. *J Biol Chem* 261: (24) 11131–11137

41 Cruse I, MD Maines (1988) Evidence suggesting that the two forms of heme oxygenase are products of different genes. *J Biol Chem* 263: (7) 3348–3353

42 Muller RM, H Taguchi, S Shibahara (1987) Nucleotide sequence and organization of the rat heme oxygenase gene. *J Biol Chem* 262: (14) 6795–6802

43 McCoubrey WK Jr, JF Ewing, MD Maines (1992) Human heme oxygenase-2: characterization and expression of a full-length cDNA and evidence suggesting that the two HO-2 transcripts may differ by choice of polyadenylation signal. *Arch Biochem Biophys* 295: (1) 13–20

44 Trakshel GM, MD Maines (1988) Detection of two heme oxygenase isoforms in the human testis. *Biochem Biophys Res Commun* 154: (1) 285–291

45 McCoubrey WK Jr, MD Maines (1994) The structure, organization and differential expression of the gene encoding rat heme oxygenase-2, *Gene* 139: (2) 155–161

46 Sun Y, MO Rotenberg, MD Maines (1990) Developmental expression of heme oxygenase isozymes in rat brain. Two HO-2 mRNAs are detected. *J Biol Chem* 265: (14) 8212–8217

47 Rotenberg MO, MD Maines (1991) Characterization of a cDNA-encoding rabbit brain heme oxygenase-2 and identification of a conserved domain among mammalian heme oxygenase isozymes: possible heme-binding site? *Arch Biochem Biophys* 290: (2) 336–344

48 Wilks A, PR Ortiz de Montellano, J Sun, TM Loehr (1996) Heme oxygenase (HO-1): His-132 stabilizes a distal water ligand and assists catalysis. *Biochemistry* 35: (3) 930–936

49 McCoubrey WK Jr, TJ Huang, MD Maines (1997) Isolation and characterization of a cDNA from the rat brain that encodes hemoprotein heme oxygenase-3. *Eur J Biochem* 247: (2) 725–732

50 Shibahara S, RM Muller, H Taguchi (1987) Transcriptional control of rat heme oxygenase by heat shock. *J Biol Chem* 262: (27) 12889–12892

51 Nascimento AL, P Luscher, RM Tyrrell (1993) Ultraviolet A (320–380 nm) radiation causes an alteration in the binding of a specific protein/protein complex to a short region of the promoter of the human heme oxygenase 1 gene. *Nucleic Acids Res* 21: (5) 1103–1109

52 Takeda K, H Fujita, S Shibahara (1995) Differential control of the metal-mediated activation of the human heme oxygenase-1 and metallothionein IIA genes. *Biochem Biophys Res Commun* 207: (1) 160–167

53 Kurata S, H Nakajima (1990) Transcriptional activation of the heme oxygenase gene by TPA in mouse M1 cells during their differentiation to macrophage. *Exp Cell Res* 191: (1) 89–94

54 Lavrovsky Y, ML Schwartzman, RD Levere, A Kappas, NG Abraham (1994) Identification of binding sites for transcription factors NF-kappa B and AP-2 in the promoter region of the human heme oxygenase 1 gene. *Proc Natl Acad Sci USA* 91: (13) 5987–5991

55 Koizumi T, N Odani, T Okuyama, A Ichikawa, M Negishi (1995) Identification of a cis-regulatory element for delta 12-prostaglandin J2-induced expression of the rat heme oxygenase gene. *J Biol Chem* 270: (37) 21779–21784

56 Negishi M, N Odani, T Koizumi, S Takahashi, A Ichikawa (1995) Involvement of protein kinase in delta 12-prostaglandin J2-induced expression of rat heme oxygenase-1 gene. *FEBS Lett* 372: (2–3) 279–282

57 Mitani K, H Fujita, A Kappas, S Sassa (1992) Heme oxygenase is a positive acute-phase reactant in human Hep3B hepatoma cells. *Blood* 79: (5) 1255–1259

58 Lee PJ, BH Jiang, BY Chin, NV Iyer, J Alam, GL Semenza, AM Choi (1997) Hypox-

ia-inducible factor-1 mediates transcriptional activation of the heme oxygenase-1 gene in response to hypoxia. *J Biol Chem* 272: (9) 5375–5381

59   Wenger RH, M Gassmann (1997) Oxygen(es) and the hypoxia-inducible factor-1. *Biol Chem* 378: (7) 609–616

60   Raju VS, WK McCoubrey Jr, MD Maines (1997) Regulation of heme oxygenase-2 by glucocorticoids in neonatal rat brain: characterization of a functional glucocorticoid response element. *Biochim Biophys Acta* 1351: (1–2) 89–104

61   Feige U, RI Morimoto, I Yahara, BS Polla (1996) *Stress-inducible cellular responses.* Birkhäuser Verlag, Basel, Boston, Berlin

62   Okinaga S, S Shibahara (1993) Identification of a nuclear protein that constitutively recognizes the sequence containing a heat-shock element. Its binding properties and possible function modulating heat-shock induction of the rat heme oxygenase gene. Eur *J Biochem* 212: (1) 167–175

63   Taketani S, H Kohno, T Yoshinaga, R Tokunaga (1988) Induction of heme oxygenase in rat hepatoma cells by exposure to heavy metals and hyperthermia. *Biochem Int* 17: (4) 665–672

64   Keyse SM RM Tyrrell (1989) Heme oxygenase is the major 32-kDa stress protein induced in human skin fibroblasts by UVA radiation, hydrogen peroxide, and sodium arsenite. *Proc Natl Acad Sci USA* 86: (1) 99–103

65   Taketani S, H Kohno, T Yoshinaga, R Tokunaga (1989) The human 32-kDa stress protein induced by exposure to arsenite and cadmium ions is heme oxygenase. *FEBS Lett* 245: (1–2) 173–176

66   Clerget M, BS Polla (1990) Erythrophagocytosis induces heat shock protein synthesis by human monocytes-macrophages. *Proc Natl Acad Sci USA* 87: (3) 1081–1085

67   Taketani S, H Sato, T Yoshinaga, R Tokunaga, T Ishii, S Bannai (1990) Induction in mouse peritoneal macrophages of 34 kDa stress protein and heme oxygenase by sulfhydryl-reactive agents. *J Biochem* (Tokyo) 108: (1) 28–32

68   Murphy BJ, KR Laderoute, SM Short, RM Sutherland (1991) The identification of heme oxygenase as a major hypoxic stress protein in Chinese hamster ovary cells. *Br J Cancer* 64: (1) 69–73

69   Okinaga S, K Takahashi, K Takeda, M Yoshizawa, H Fujita, H Sasaki, S Shibahara (1996) Regulation of human heme oxygenase-1 gene expression under thermal stress. *Blood* 87: (12) 5074–5084

70   Cantoni L, M Carelli, P Ghezzi, R Delgado, R Faggioni, M Rizzardini (1995) Mechanisms of interleukin-2-induced depression of hepatic cytochrome P-450 in mice. *Eur J Pharmacol* 292: (3–4) 257–263

71   Cantoni L, C Rossi, M Rizzardini, M Gadina, P Ghezzi (1991) Interleukin-1 and tumour necrosis factor induce hepatic haem oxygenase Feedback regulation by glucocorticoids. *Biochem J* 279: (Pt 3) 891–894

72   Rizzardini M, M Terao, F Falciani, L Cantoni (1993) Cytokine induction of haem oxygenase mRNA in mouse liver. Interleukin 1 transcriptionally activates the haem oxygenase gene. *Biochem J* 290: (Pt 2) 343–347

73 Willis D, AR Moore, R Frederick, DA Willoughby (1996) Heme oxygenase: a novel target for the modulation of the inflammatory response. *Nature Med* 2: (1) 87–90

74 Willis D (1995) Expression and modulatory effects of heme oxygenase in acute inflammation in the rat. *Inflamm Res* 44 Suppl 2: S218–S220

75 Moore AR, D Willis, D Gilroy, A Tomlinson, I Appleton, DA Willoughby (1995) Cyclooxygenase in rat pleural hypersensitivity reactions. *Adv Prostaglandin Thromboxane Leukot Res* 23: 349–351

76 Polla BS, S Kantengwa, GJ Gleich, M Kondo, CM Reimert, AF Junod (1993) Spontaneous heat shock protein synthesis by alveolar macrophages in interstitial lung disease associated with phagocytosis of eosinophils. *Eur Respir J* 6: (4) 483–488

77 Laniado-Schwartzman M, NG Abraham, M Conners, MW Dunn, RD Levere, A Kappas (1997) Heme oxygenase induction with attenuation of experimentally induced corneal inflammation. *Biochem Pharmacol* 53: (8) 1069–1075

78 Mosley K, DE Wembridge, V Cattell, HT Cook (1998) Heme oxygenase is induced in nephrotoxic nephritis and hemin, a stimulator of heme oxygenase synthesis, ameliorates disease. *Kidney Int* 53: (3) 672–678

79 Vogt BA, TP Shanley, A Croatt, J Alam, KJ Johnson, KA Nath (1996) Glomerular inflammation induces resistance to tubular injury in the rat. A novel form of acquired, heme oxygenase-dependent resistance to renal injury. *J Clin Invest* 98: (9) 2139–2145

80 Vogt BA, J Alam, AJ Croatt, GM Vercellotti, KA Nath (1995) Acquired resistance to acute oxidative stress Possible role of heme oxygenase and ferritin. *Lab Invest* 72: (4) 474–483

81 Nath KA, G Balla, GM Vercellotti, J Balla, HS Jacob, MD Levitt, ME Rosenberg (1992) Induction of heme oxygenase is a rapid, protective response in rhabdomyolysis in the rat. *J Clin Invest* 90: (1) 267–270

82 Agarwal A, J Balla, J Alam, AJ Croatt, KA Nath (1995) Induction of heme oxygenase in toxic renal injury: a protective role in cisplatin nephrotoxicity in the rat. *Kidney Int* 48: (4) 1298–1307

83 Paller MS, KA Nath, ME Rosenberg (1993) Heme oxygenase is not expressed as a stress protein after renal ischemia. *J Lab Clin Med* 122: (3) 341–345

84 Choi AM, K Knobil, SL Otterbein, DA Eastman, DB Jacoby (1996) Oxidant stress responses in influenza virus pneumonia: gene expression and transcription factor activation. *Am J Physiol* 271: (3 Pt 1) L383–L391

85 Takahashi Y, S Takahashi, T Yoshimi, T Miura, K Mochitate, T Kobayashi (1997) Increases in the mRNA levels of gamma-glutamyltransferase and heme oxygenase-1 in the rat lung after ozone exposure. *Biochem Pharmacol* 53: (7) 1061–1064

86 Liao F, A Andalibi, JH Qiao, H Allayee, AM Fogelman, AJ Lusis (1994) Genetic evidence for a common pathway mediating oxidative stress, inflammatory gene induction, and aortic fatty streak formation in mice. *J Clin Invest* 94: (2) 877–884

87 Liao F, A Andalibi, FC deBeer, AM Fogelman, AJ Lusis (1993) Genetic control of inflammatory gene induction and NF-kappa B-like transcription factor activation in response to an atherogenic diet in mice. *J Clin Invest* 91: (6) 2572–2579

88 Toda A, T Kihara, N Ono, A Nagamatsu, H Shimeno (1996) Liver haem metabolism in adjuvant-induced arthritic rats. *Xenobiotica* 26: (4) 415–423

89 Poss KD, S Tonegawa (1997) Heme oxygenase 1 is required for mammalian iron reutilization. *Proc Natl Acad Sci USA* 94: (20) 10919–10924

90 Poss KD, S Tonegawa (1997) Reduced stress defense in heme oxygenase 1-deficient cells. *Proc Natl Acad Sci USA* 94: (20) 10925–10930

91 Poss KD, MJ Thomas, AK Ebralidze, TJ O'Dell, S Tonegawa (1995) Hippocampal long-term potentiation is normal in heme oxygenase-2 mutant mice. *Neuron* 15: (4) 867–873

92 Balla J, HS Jacob, G Balla, K Nath, JW Eaton, GM Vercellotti (1993) Endothelial-cell heme uptake from heme proteins: induction of sensitization and desensitization to oxidant damage. *Proc Natl Acad Sci USA* 90: (20) 9285–9289

93 Neuzil J, R Stocker (1993) Bilirubin attenuates radical-mediated damage to serum albumin. *FEBS Lett* 331: (3) 281–284

94 Stocker R, BN Ames (1987) Potential role of conjugated bilirubin and copper in the metabolism of lipid peroxides in bile. *Proc Natl Acad Sci USA* 84: (22) 8130–8134

95 Stocker R, Y Yamamoto, AF McDonagh, AN Glazer, BN Ames (1987) Bilirubin is an antioxidant of possible physiological importance. *Science* 235: (4792) 1043–1046

96 Stocker R, AN Glazer, BN Ames (1987) Antioxidant activity of albumin-bound bilirubin. *Proc Natl Acad Sci USA* 84: (16) 5918–5922

97 Wu TW, J Wu, RK Li, D Mickle, D Carey (1991) Albumin-bound bilirubins protect human ventricular myocytes against oxyradical damage. *Biochem Cell Biol* 69: (10–11) 683–688

98 Stocker R, E Peterhans (1989) Synergistic interaction between vitamin E and the bile pigments bilirubin and biliverdin. *Biochim Biophys Acta* 1002: (2) 238–244

99 Stocker R, E Peterhans (1989) Antioxidant properties of conjugated bilirubin and biliverdin: biologically relevant scavenging of hypochlorous acid. *Free Radic Res Commun* 6: (1) 57–66

100 Oyanagui Y (1997) Natural antioxidants enhance and prolong the oxyradical/NO-related suppression by dexamethasone of ischemic and histamine paw edema in mice. Inflammation 21: (6) 643–654

101 Makino Y, K Okamoto, N Yoshikawa, M Aoshima, K Hirota, J Yodoi, K Umesono, I Makino, H Tanaka (1996) Thioredoxin: a redox-regulating cellular cofactor for glucocorticoid hormone action. Cross talk between endocrine control of stress response and cellular antioxidant defense system. *J Clin Invest* 98: (11) 2469–2477

102 Ishikawa K, M Navab, N Leitinger, AM Fogelman, AJ Lusis (1997) Induction of heme oxygenase-1 inhibits the monocyte transmigration induced by mildly oxidized LDL. *J Clin Invest* 100: (5) 1209–1216

103 Nakagami T, K Toyomura, T Kinoshita, S Morisawa (1993) A beneficial role of bile pigments as an endogenous tissue protector: anti-complement effects of biliverdin and conjugated bilirubin. *Biochim Biophys Acta* 1158: (2) 189–193

104 Haga Y, MA Tempero, RK Zetterman (1996) Unconjugated bilirubin inhibits *in vitro*

cytotoxic T lymphocyte activity of human lymphocytes. *Biochim Biophys Acta* 1317: (1) 65–70

105 Haga Y, MA Tempero, RK Zetterman (1996) Unconjugated bilirubin inhibits *in vitro* major histocompatibility complex-unrestricted cytotoxicity of human lymphocytes. *Biochim Biophys Acta* 1316: (1) 29–34

106 Haga Y, MA Tempero, D Kay, RK Zetterman (1996) Intracellular accumulation of unconjugated bilirubin inhibits phytohemagglutin-induced proliferation and inter-leukin-2 production of human lymphocytes. *Dig Dis Sci* 41: (7) 1468–1474

107 Conners MS, RA Stoltz, KL Davis, MW Dunn, NG Abraham, RD Levere, M Lania-do-Schwartzman (1995) A closed eye contact lens model of corneal inflammation. Part 2: Inhibition of cytochrome P450 arachidonic acid metabolism alleviates inflammatory sequelae. *Invest Ophthalmol Vis Sci* 36: (5) 841–850

108 Martasek P, ML Schwartzman, AI Goodman, KB Solangi, RD Levere, NG Abraham (1991) Hemin and L-arginine regulation of blood pressure in spontaneous hypertensive rats. *J Am Soc Nephrol* 2: (6) 1078–1084

109 Willis D, A Tomlinson, R Frederick, MJ Paul-Clark, DA Willoughby (1995) Modulation of heme oxygenase activity in rat brain and spleen by inhibitors and donors of nitric oxide. *Biochem Biophys Res Commun* 214: (3) 1152–1156

110 Marks GS, JF Brien, K Nakatsu, BE McLaughlin (1991) Does carbon monoxide have a physiological function? *Trends Pharmacol Sci* 12: (5) 185–188

111 Ingi T, J Cheng, GV Ronnett (1996) Carbon monoxide: an endogenous modulator of the nitric oxide-cyclic GMP signaling system. *Neuron* 16: (4) 835–842

112 Jiang C, AT Ting, B Seed (1998) PPAR-gamma agonists inhibit production of monocyte inflammatory cytokines. *Nature* 391: (6662) 82–86

113 Ricote M, AC Li, TM Willson, CJ Kelly, CK Glass (1998) The peroxisome prolifera-tor-activated receptor-gamma is a negative regulator of macrophage activation. *Nature* 391: (6662) 79–82

114 Rossi A, G Elia, MG Santoro (1997) Inhibition of nuclear factor kappa B by prosta-glandin A1: an effect associated with heat shock transcription factor activation. *Proc Natl Acad Sci USA* 94: (2) 746–750

115 Silver BJ, BD Hamilton, Z Toossi (1997) Suppression of TNF-alpha gene expression by hemin: implications for the role of iron homeostasis in host inflammatory responses. *J Leukoc Biol* 62: (4) 547–552

116 Balla J, HS Jacob, G Balla, K Nath, GM Vercellotti (1992) Endothelial cell heme oxy-genase and ferritin induction by heme proteins: a possible mechanism limiting shock damage. *Trans Assoc Am Physicians* 105: 1–6

117 Harada T, M Baba, I Torii, S Morikawa (1987) Ferritin selectively suppresses delayed-type hypersensitivity responses at induction or effector phase. *Cell Immunol* 109: (1) 75–88

118 Vile GF, S Basu-Modak, C Waltner, RM Tyrrell (1994) Heme oxygenase 1 mediates an adaptive response to oxidative stress in human skin fibroblasts. *Proc Natl Acad Sci USA* 91: (7) 2607–2610

119 Vile GF, RM Tyrrell (1993) Oxidative stress resulting from ultraviolet A irradiation of human skin fibroblasts leads to a heme oxygenase-dependent increase in ferritin. *J Biol Chem* 268: (20) 14678–14681

120 Dennery PA, KJ Sridhar, CS Lee, HE Wong, V Shokoohi, PA Rodgers, DR Spitz (1997) Heme oxygenase-mediated resistance to oxygen toxicity in hamster fibroblasts. *J Biol Chem* 272: (23) 14937–14942

121 Dennery PA, HE Wong, KJ Sridhar, PA Rodgers, JE Sim, DR Spitz (1996) Differences in basal and hyperoxia-associated HO expression in oxidant-resistant hamster fibroblasts. *Am J Physiol* 271: (4 Pt 1) L672–L679

122 Otterbein L, BY Chin, SL Otterbein, VC Lowe, HE Fessler, AM Choi (1997) Mechanism of hemoglobin-induced protection against endotoxemia in rats: a ferritin-independent pathway. *Am J Physiol* 272: (2 Pt 1) L268–L275

123 Nutter LM, EE Sierra, EO Ngo (1994) Heme oxygenase does not protect human cells against oxidant stress. *J Lab Clin Med* 123: (4) 506-514

124 Da Silva JL, T Morishita, B Escalante, R Staudinger, G Drummond, MS Goligorsky, JD Lutton, NG Abraham (1996) Dual role of heme oxygenase in epithelial cell injury: contrasting effects of short-term and long-term exposure to oxidant stress. *J Lab Clin Med* 128: (3) 290–296

125 Dennery PA, DR Spitz, G Yang, A Tatarov, CS Lee, ML Shegog, KD Poss (1998) Oxygen toxicity and iron accumulation in the lungs of mice lacking heme oxygenase-2. *J Clin Invest* 101: (5) 1001–1011

126 McCoubrey WK Jr, TJ Huang, MD Maines (1997) Heme oxygenase-2 is a hemoprotein and binds heme through heme regulatory motifs that are not involved in heme catalysis. *J Biol Chem* 272: (19) 12568–12574

127 Lander HM (1997) An essential role for free radicals and derived species in signal transduction. *FASEB J* 11: (2) 118–124

128 Winyard PG, DR Blake (1997) Antioxidants, redox-regulated transcription factors, and inflammation. *Adv Pharmacol* 38: 403–421

129 Sen CK, L Packer (1996) Antioxidant and redox regulation of gene transcription. *FASEB J* 10: (7) 709–720

130 Piette J, B Piret, G Bonizzi, S Schoonbroodt, MP Merville, S Legrand-Poels, V Bours (1997) Multiple redox regulation in NF-kappaB transcription factor activation. *Biol Chem* 378: (11) 1237–1245

131 Parrillo JE (1993) Pathogenetic mechanisms of septic shock. *N Engl J Med* 328: (20) 1471–1477

132 Camhi SL, J Alam, L Otterbein, SL Sylvester, AM Choi (1995) Induction of heme oxygenase-1 gene expression by lipopolysaccharide is mediated by AP-1 activation. *Am J Respir Cell Mol Biol* 13: (4) 387–398

133 Kurata S, M Matsumoto, H Nakajima (1996) Transcriptional control of the heme oxygenase gene in mouse M1 cells during their TPA-induced differentiation into macrophages. *J Cell Biochem* 62: (3) 314–324

134 Morita T, MA Perrella, ME Lee, S Kourembanas (1995) Smooth muscle cell-derived car-

bon monoxide is a regulator of vascular cGMP. *Proc Natl Acad Sci USA* 92: (5) 1475–1479

135 Morita T, S Kourembanas (1995) Endothelial cell expression of vasoconstrictors and growth factors is regulated by smooth muscle cell-derived carbon monoxide. *J Clin Invest* 96: (6) 2676–2682

136 Otterbein L, SL Sylvester, AM Choi (1995) Hemoglobin provides protection against lethal endotoxemia in rats: the role of heme oxygenase-1. *Am J Respir Cell Mol Biol* 13: (5) 595–601

137 Yet SF, A Pellacani, C Patterson, L Tan, SC Folta, L Foster, WS Lee, CM Hsieh, MA Perrella (1997) Induction of heme oxygenase-1 expression in vascular smooth muscle cells. A link to endotoxic shock. *J Biol Chem* 272: (7) 4295–4301

138 Johnson RA, M Lavesa, K DeSeyn, J Scholer, A Nasjletti (1996) Heme oxygenase substrates acutely lower blood pressure in hypertensive rats. *Am J Physiol* 271: (3 Pt 2) H1132–H1138

139 Johnson RA, M Lavesa, B Askari, NG Abraham, A Nasjletti (1995) A heme oxygenase product, presumably carbon monoxide, mediates a vasodepressor function in rats. *Hypertension* 25: (2) 166–169

140 Katayose D, S Isoyama, H Fujita, S Shibahara (1993) Separate regulation of heme oxygenase and heat shock protein 70 mRNA expression in the rat heart by hemodynamic stress. *Biochem Biophys Res Commun* 191: (2) 587–594

141 Gardiner SM, PA Kemp, B Fallgren, T Bennett (1994) Effects of chronic infusions of alpha-trinositol on regional and cardiac haemodynamics in conscious rats. *Br J Pharmacol* 113: (1) 129–136

142 Tomlinson A, SM Gardiner, D Willis, J Ali, PA Kemp, DA Willoughby, T Bennett (1998) Temporal and spatial expresssion of the inducible isoforms of cyclooxygenase, nitric oxide synthase and heme oxygenase in tissues from rats infused with LPS in the conscious state. *Br J Pharmacol* 123: 178P

143 Sacerdoti D, B Escalante, NG Abraham, JC McGiff, RD Levere, ML Schwartzman (1989) Treatment with tin prevents the development of hypertension in spontaneously hypertensive rats. *Science* 243: (4889) 388–390

144 Vercellotti GM, JP Tolins (1993) Endothelial activation and the kidney: vasomediator modulation and antioxidant strategies. *Am J Kidney Dis* 21: (3) 331–343

145 Agarwal A, Y Kim, AJ Matas, J Alam, KA Nath (1996) Gas-generating systems in acute renal allograft rejection in the rat. Co-induction of heme oxygenase and nitric oxide synthase. *Transplantation* 61: (1) 93–98

146 Bach FH, C Ferran, P Hechenleitner, W Mark, N Koyamada, T Miyatake, H Winkler, A Badrichani, D Candinas, WW Hancock (1997) Accommodation of vascularized xenografts: expression of "protective genes" by donor endothelial cells in a host Th2 cytokine environment. *Nature Med* 3: (2) 196–204

147 Buelow R, WJ Burlingham, C Clayberger (1995) Immunomodulation by soluble HLA class I. *Transplantation* 59: (5) 649–654

148 Cuturi MC, R Josien, P Douillard, C Pannetier, D Cantarovich, H Smit, S Menoret, P

Pouletty, C Clayberger, JP Soulillou (1995) Prolongation of allogeneic heart graft survival in rats by administration of a peptide (a.a. 75-84) from the alpha 1 helix of the first domain of HLA-B7 01. *Transplantation* 59: (5) 661–669

149 Nisco S, P Vriens, G Hoyt, SC Lyu, F Farfan, P Pouletty, AM Krensky, C Clayberger (1994) Induction of allograft tolerance in rats by an HLA class-I-derived peptide and cyclosporine A. *J Immunol* 152: (8) 3786–3792

150 Woo J, L Gao, MC Cornejo, R Buelow (1995) A synthetic dimeric HLA class I peptide inhibits T cell activity *in vitro* and prolongs allogeneic heart graft survival in a mouse model. *Transplantation* 60: (10) 1156–1163

151 Iyer S, J Woo, MC Cornejo, L Gao, WMcCoubrey, M Maines, R Buelow (1998) Characterization and biological significance of immunosuppressive peptide D270275-84(E → V) binding protein. Isolation of heme oxygenase-1. *J Biol Chem* 273: (5) 2692–2697

152 Raju VS, MD Maines (1996) Renal ischemia/reperfusion up-regulates heme oxygenase-1 (HSP32) expression and increases cGMP in rat heart. *J Pharmacol Exp Ther* 277: (3) 1814–1822

153 Maulik N, DT Engelman, M Watanabe, RM Engelman, DK Das (1996) Nitric oxide – a retrograde messenger for carbon monoxide signaling in ischemic heart. *Mol Cell Biochem* 157: (1–2) 75–86

154 Maulik N, DT Engelman, M Watanabe, RM Engelman, JA Rousou, JE Flack 3rd, DW Deaton, NV Gorbunov, NM Elsayed, VE Kagan, DK Das (1996) Nitric oxide/carbon monoxide. A molecular switch for myocardial preservation during ischemia. *Circulation* 94: (9 Suppl) II 398–406

155 Maulik N, HS Sharma, DK Das (1996) Induction of the haem oxygenase gene expression during the reperfusion of ischemic rat myocardium. *J Mol Cell Cardiol* 28: (6) 1261–1270

156 Sharma HS, N Maulik, BC Gho, DK Das, PD Verdouw (1996) Coordinated expression of heme oxygenase-1 and ubiquitin in the porcine heart subjected to ischemia and reperfusion. *Mol Cell Biochem* 157: (1-2) 111–116

157 Bergeron M, DM Ferriero, HJ Vreman, DK Stevenson, FR Sharp (1997) Hypoxia-ischemia, but not hypoxia alone, induces the expression of heme oxygenase-1 (HSP32) in newborn rat brain. *J Cereb Blood Flow Metab* 17: (6) 647–658

158 Geddes JW, LC Pettigrew, ML Holtz, SD Craddock, MD Maines (1996) Permanent focal and transient global cerebral ischemia increase glial and neuronal expression of heme oxygenase-1, but not heme oxygenase-2, protein in rat brain. *Neurosci Lett* 210: (3) 205–208

159 Nimura T, PR Weinstein, SM Massa, S Panter, FR Sharp (1996) Heme oxygenase-1 (HO-1) protein induction in rat brain following focal ischemia. *Brain Res Mol Brain Res* 37: (1–2) 201–208

160 Paschen W, A Uto, B Djuricic, J Schmitt (1994) Hemeoxygenase expression after reversible ischemia of rat brain. *Neurosci Lett* 180: (1) 5–8

161 Takeda A, T Kimpara, H Onodera, Y Itoyama, S Shibahara, K Kogure (1996) Region-

al difference in induction of heme oxygenase-1 protein following rat transient forebrain ischemia. *Neurosci Lett* 205: (3) 169–172

162 Takeda A, H Onodera, A Sugimoto, Y Itoyama, K Kogure, S Shibahara (1994) Increased expression of heme oxygenase mRNA in rat brain following transient forebrain ischemia. *Brain Res* 666: (1) 120–124

163 Hoshida S, K Aoki, M Nishida, N Yamashita, J Igarashi, M Hori, T Kuzuya, M Tada (1997) Effects of preconditioning with ebselen on glutathione metabolism and stress protein expression. *J Pharmacol Exp Ther* 281: (3) 1471–1475

164 Kilgore KS, BR Lucchesi (1993) Reperfusion injury after myocardial infarction: the role of free radicals and the inflammatory response. *Clin Biochem* 26: (5) 359-370

165 Lucchesi BR (1994) Complement, neutrophils and free radicals: mediators of reperfusion injury. *Arzneimittelforschung* 44: (3A) 420–432

166 Goodman AI, M Choudhury, JL Da Silva, ML Schwartzman, NG Abraham (1997) Overexpression of the heme oxygenase gene in renal cell carcinoma. *Proc Soc Exp Biol Med* 214: (1) 54–61

167 Goodman AI, M Choudhury, JL Da Silva, S Jiang, NG Abraham (1996) Quantitative measurement of heme oxygenase-1 in the human renal adenocarcinoma. *J Cell Biochem* 63: (3) 342–348

168 Maines MD, PA Abrahamsson (1996) Expression of heme oxygenase-1 (HSP32) in human prostate: normal, hyperplastic, and tumor tissue distribution. *Urology* 47: (5) 727–733

169 Helqvist S, B Sehested Hansen, J Johannesen, H Ullits Andersen, J Hoiriis Nielsen, J Nerup (1989) Interleukin 1 induces new protein formation in isolated rat islets of Langerhans. *Acta Endocrinol (Copenh)* 121: (1) 136–140

170 Neil TK, RA Stoltz, S Jiang, M Laniado-Schwartzman, MW Dunn, RD Levere, A Kappas, NG Abraham (1995) Modulation of corneal heme oxygenase expression by oxidative stress agents. *Ocul Pharmacol Ther* 11: (3) 455–468

171 Terry CM, JA Clikeman, JR Hoidal, KS Callahan (1998) Effect of tumor necrosis factor-alpha and interleukin-1alpha on heme oxygenase-1 expression in human endothelial cells. *Am J Physiol* 274: (3) H883–H891

172 Tetsuka T, D Daphna-Iken, SK Srivastava, AR Morrison (1995) Regulation of heme oxygenase mRNA in mesangial cells: prostaglandin E2 negatively modulates interleukin-1-induced heme oxygenase-1 mRNA. *Biochem Biophys Res Commun* 212: (2) 617–623

173 Helqvist S, BS Polla, J Johannesen, J Nerup (1991) Heat shock protein induction in rat pancreatic islets by recombinant human interleukin 1 beta. *Diabetologia* 34: (3) 150–156

174 Strandell E, K Buschard, J Saldeen, N Welsh (1995) Interleukin-1 beta induces the expression of hsp70, heme oxygenase and Mn-SOD in FACS-purified rat islet beta-cells, but not in alpha-cells. *Immunol Lett* 48: (2) 145–148

175 Fukuda Y, S Sassa (1994) Suppression of cytochrome P450IA1 by interleukin-6 in human HepG2 hepatoma cells. *Biochem Pharmacol* 47: (7) 1187–1195

176 Fukuda Y, S Sassa (1993) Effect of interleukin-11 on the levels of mRNAs encoding heme oxygenase and haptoglobin in human HepG2 hepatoma cells. *Biochem Biophys Res Commun* 193: (1) 297–302

177 Mitani K, H Fujita, S Sassa, A Kappas (1991) A heat-inducible nuclear factor that binds to the heat-shock element of the human haem oxygenase gene. *Biochem J* 277: (Pt 3) 895–897

178 Mitani K, H Fujita, S Sassa, A Kappas (1990) Activation of heme oxygenase and heat shock protein 70 genes by stress in human hepatoma cells. *Biochem Biophys Res Commun* 166: (3) 1429–1434

179 Mitani K, H Fujita, S Sassa, A Kappas (1989) Heat shock induction of heme oxygenase mRNA in human Hep 3B hepatoma cells. *Biochem Biophys Res Commun* 165: (1) 437–441

180 Wagener E, E Feldman, T de Witte, NG Abraham (1997) Heme induces the expression of adhesion molecules ICAM-1, VCAM-1, and E selectin in vascular endothelial cells. *Proc Soc Exp Biol Med* 216: (3) 456–463

181 Kutty RK, CN Nagineni, G Kutty, JJ Hooks, GJ Chader, B Wiggert (1994) Increased expression of heme oxygenase-1 in human retinal pigment epithelial cells by transforming growth factor-beta. *J Cell Physiol* 159: (2) 371–378

182 Kurata S, M Matsumoto, U Yamashita (1996) Concomitant transcriptional activation of nitric oxide synthase and heme oxygenase genes during nitric oxide-mediated macrophage cytostasis. *J Biochem (Tokyo)* 120: (1) 49–52

183 Motterlini R, R Foresti, M Intaglietta, RM Winslow (1996) NO-mediated activation of heme oxygenase: endogenous cytoprotection against oxidative stress to endothelium. *Am J Physiol* 270: (1 Pt 2) H107–H114

184 Lutton JD, JL Da Silva, S Moqattash, AC Brown, RD Levere, NG Abraham (1992) Differential induction of heme oxygenase in the hepatocarcinoma cell line (Hep3B) by environmental agents. *J Cell Biochem* 49: (3) 259–265

185 Kitamura Y, Y Matsuoka, Y Nomura, T Taniguchi (1998) Induction of inducible nitric oxide synthase and heme oxygenase-1 in rat glial cells. *Life Sci* 62: (17–18) 1717–1721

186 Kurata S, M Matsumoto, Y Tsuji, H Nakajima (1996) Lipopolysaccharide activates transcription of the heme oxygenase gene in mouse M1 cells through oxidative activation of nuclear factor kappa B. *Eur J Biochem* 239: (3) 566–571

187 Durante W, MH Kroll, N Christodoulides, KJ Peyton, AI Schafer (1997) Nitric oxide induces heme oxygenase-1 gene expression and carbon monoxide production in vascular smooth muscle cells. *Circ Res* 80: (4) 557–564

188 Foresti R, JE Clark, CJ Green, R Motterlini (1997) Thiol compounds interact with nitric oxide in regulating heme oxygenase-1 induction in endothelial cells Involvement of superoxide and peroxynitrite anions. *J Biol Chem* 272: (29) 18411–18417

189 Hara E, K Takahashi, T Tominaga, T Kumabe, T Kayama, H Suzuki, H Fujita, T Yoshimoto, K Shirato, S Shibahara (1996) Expression of heme oxygenase and inducible nitric oxide synthase mRNA in human brain tumors. *Biochem Biophys Res Commun* 224: (1) 153–158

190 Polte T, S Oberle, H Schroeder (1997) The nitric oxide donor SIN-1 protects endothelial cells from tumor necrosis factor-alpha mediated cytotoxicity: Possible role for cyclic GMP and heme oxygenase. *J Mol Cell Cardiol* 29: (12) 3305–3310

191 Yee EL, BR Pitt, TR Billiar, YM Kim (1996) Effect of nitric oxide on heme metabolism in pulmonary artery endothelial cells. *Am J Physiol* 271: (4 Pt 1) L512–L518

192 Hartsfield CL, J Alam, JL Cook, AM Choi (1997) Regulation of heme oxygenase-1 gene expression in vascular smooth muscle cells by nitric oxide. *Am J Physiol* 273: (5 Pt 1) L980–L988

193 Sandau K, J Pfeilschifter, B Bruene (1998) Nitrosactive and oxidative stress induced heme oxygenase-1 accumulation in rat mesangial cells. *Eur J Pharmacol* 342: (1) 77–84

194 Sato H, RC Siow, S Bartlett, S Taketani, T Ishii, S Bannai, GE Mann (1997) Expression of stress proteins heme oxygenase-1 and -2 in acute pancreatitis and pancreatic islet betaTC3 and acinar AR42J cells. *FEBS Lett* 405: (2) 219–223

195 Steels EL, K Watson, PG Parsons (1992) Relationships between thermotolerance, oxidative stress responses and induction of stress proteins in human tumour cell lines. *Biochem Pharmacol* 44: (11) 2123–2129

196 Nose K, M Shibanuma, K Kikuchi, H Kageyama, S Sakiyama, T Kuroki (1991) Transcriptional activation of early-response genes by hydrogen peroxide in a mouse osteoblastic cell line. *Eur J Biochem* 201: (1) 99–106

197 Applegate LA, A Noel, G Vile, E Frenk, RM Tyrrell (1995) Two genes contribute to different extents to the heme oxygenase enzyme activity measured in cultured human skin fibroblasts and keratinocytes: implications for protection against oxidant stress. *Photochem Photobiol* 61: (3) 285–291

198 Applegate LA, P Luscher, RM Tyrrell (1991) Induction of heme oxygenase: a general response to oxidant stress in cultured mammalian cells. *Cancer Res* 51: (3) 974–978

199 Lautier D, P Luscher, RM Tyrrell (1992) Endogenous glutathione levels modulate both constitutive and UVA radiation/hydrogen peroxide inducible expression of the human heme oxygenase gene. *Carcinogenesis* 13: (2) 227–232

200 Borger DR, DA Essig (1998) Induction of HSP 32 gene in hypoxic cardiomyocytes is attenuated by treatment with N-acetyl-L-cysteine. *Am J Physiol* 274: (3) H965–H973

201 Hoshida S, M Nishida, N Yamashita, J Igarashi, K Aoki, M Hori, T Kuzuya, M Tada (1996) Heme oxygenase-1 expression and its relation to oxidative stress during primary culture of cardiomyocytes. *J Mol Cell Cardiol* 28: (9) 1845–1855

202 Sato H, T Ishii, Y Sugita, N Tateishi, S Bannai (1993) Induction of a 23 kDa stress protein by oxidative and sulfhydryl-reactive agents in mouse peritoneal macrophages. *Biochim Biophys Acta* 1148: (1) 127–132

203 Rossi A, MG Santoro (1995) Induction by prostaglandin A1 of haem oxygenase in myoblastic cells: an effect independent of expression of the 70 kDa heat shock protein. *Biochem J* 308: (Pt 2) 455–463

204 Choi AM, RW Tucker, SG Carlson, G Weigand, NJ Holbrook (1994) Calcium mediates expression of stress-response genes in prostaglandin A2-induced growth arrest. *FASEB J* 8: (13) 1048–1054

205 Koizumi T, M Negishi, A Ichikawa (1992) Inhibitory effect of an intracellular glutathione on delta 12-prostaglandin J2-induced protein syntheses in porcine aortic endothelial cells. *Biochem Pharmacol* 44: (8) 1597–1602

206 Koizumi T, M Negishi, A Ichikawa (1992) Induction of heme oxygenase by delta 12-prostaglandin J2 in porcine aortic endothelial cells. *Prostaglandins* 43: (2) 121–131

207 Agarwal A, J Balla, G Balla, AJ Croatt, GM Vercellotti, KA Nath (1996) Renal tubular epithelial cells mimic endothelial cells upon exposure to oxidized LDL. *Am J Physiol* 271: (4 Pt 2) F814–F823

208 Siow RC, T Ishii, H Sato, S Taketani, DS Leake, JH Sweiry, JD Pearson, S Bannai, GE Mann (1995) Induction of the antioxidant stress proteins heme oxygenase-1 and MSP23 by stress agents and oxidised LDL in cultured vascular smooth muscle cells. *FEBS Lett* 368: (2) 239–242

209 Yamaguchi M, H Sato, S Bannai (1993) Induction of stress proteins in mouse peritoneal macrophages by oxidized low-density lipoprotein. *Biochem Biophys Res Commun* 193: (3) 1198–1201

210 Jornot L, AF Junod (1993) Variable glutathione levels and expression of antioxidant enzymes in human endothelial cells. *Am J Physiol* 264: (5 Pt 1) L482–L489

211 Lee PJ, J Alam, SL Sylvester, N Inamdar, L Otterbein, AM Choi (1996) Regulation of heme oxygenase-1 expression *in vivo* and *in vitro* in hyperoxic lung injury. *Am J Respir Cell Mol Biol* 14: (6) 556–568

212 Visner GA, S Fogg, HS Nick (1996) Hyperoxia-responsive proteins in rat pulmonary microvascular endothelial cells. *Am J Physiol* 270: (4 Pt 1) L517–L525

213 Guyton KZ, DR Spitz, NJ Holbrook (1996) Expression of stress response genes GADD153, c-jun, and heme oxygenase-1 in H2O2- and O2-resistant fibroblasts. *Free Radic Biol Med* 20: (5) 735–741

214 Eyssen-Hernandez R, A Ladoux, C Frelin (1996) Differential regulation of cardiac heme oxygenase-1 and vascular endothelial growth factor mRNA expressions by hemin, heavy metals, heat shock and anoxia. *FEBS Lett* 382: (3) 229–233

215 Huang TJ, GM Trakshel, MD Maines (1989) Detection of 10 variants of biliverdin reductase in rat liver by two-dimensional gel electrophoresis. *J Biol Chem* 264: (14) 7844–7849

216 Kutty RK, RF Daniel, DE Ryan, W Levin, MD Maines (1988) Rat liver cytochrome P-450b, P-420b, and P-420c are degraded to biliverdin by heme oxygenase. *Arch Biochem Biophys* 260: (2) 638–644

217 Lutton JD, RD Levere, NG Abraham (1991) Physiologic role of heme and cytochrome P-450 in hematopoietic cells. *Proc Soc Exp Biol Med* 196: (3) 260–269

218 Wu TW, KP Fung, J Wu, CC Yang, RD Weiser (1996) Antioxidation of human low density lipoprotein by unconjugated and conjugated bilirubins. *Biochem Pharmacol* 51: (6) 859–862

219 Hopkins PN, LL Wu, SC Hunt, BC James, GM Vincent, RR Williams (1996) Higher serum bilirubin is associated with decreased risk for early familial coronary artery disease. *Arterioscler Throm Biol* 16: (2) 250–255

220 Schwertner HA, WG Jackson, G Tolan (1994) Association of low serum concentration of bilirubin with increased risk of coronary artery disease. *Clin Chem* 40: (1) 18–23

221 Wang LS, TS Lee, FY Lee, RC Pai, LC Chau (1998) Expression of heme oxygenase-1 in artherosclerotic lesions. *Am J Pathol* 152: (3) 711–720

222 Colville-Nash PR, SS Qureshi, D Willis, DA Willoughby (1998) Inhibition of inducible nitric oxide synthase by peroxisome proliferator-activated receptor agonists: Correlation with induction of heme oxygenase-1. *J Immunol* 161: 978–984

223 Waxman K (1996) Shock: ischemia, reperfusion, and inflammation. *New Horiz* 4: (2) 153–160

224 Tacchini L, L Shiaffonati, C Pappalardo, S Gatti, A Bernelli-Zazzera (1993) Expression of HSP 70, immediate-early response and heme oxygenase genes in ischemic-reperfused rat liver. *Lab Invest* 68: (4) 465–471

225 Yamaguchi T, M Terakado, F Horio, K Aoki, M Tanaka, H Nakajima (1996) Role of bilirubin as an antioxidant in an ischemia-reperfusion of rat liver and induction of heme oxygenase. *Biochem Biophys Res Commun* 223: (1) 129–135

226 Dennery PA, AF McDonagh, DR Spitz, PA Rodgers, DK Stevenson (1995) Hyperbilirubinemia results in reduced oxidative injury in Gunn rats exposed to hyperoxia. *Free Radic Biol Med* 19: (4) 395–404

Inducible enzymes in the pathogenesis of rheumatoid arthritis

Vivienne R. Winrow and David R. Blake

Bone and Joint Medicine, Department of Postgraduate Medicine, University of Bath, Claverton Down, Bath BA2 7AY, UK

Rheumatoid arthritis (RA) is a progressive debilitating inflammatory disease of unknown aetiology. The incidence of arthritis within the community is approximately 2–3% and, on average, accounts for the loss of 88 million working days per year. Thus, chronic inflammatory diseases like RA impose a considerable financial burden on national economies. Intervention therapy tends merely to limit inflammation and may induce adverse reactions but, in general, there is little effect on disease progression, possibly highlighting our misconception of the disease process. To this end, novel therapeutic targets are essential.

Pathophysiology of rheumatoid arthritis

The normal diarthrodial synovial joint consists of two bone ends capped with hyaline cartilage and joined by a fibrous capsule which is contiguous with the periosteum of the bones. The synovium lines the capsule and, in health, comprises an intimal lining layer 2–3 cells thick, resting on vascular sub-intimal tissue which may be adipose, fibrous or areolar. The normal joint space contains a small volume of a clear viscous liquid termed synovial fluid.

Rheumatoid arthritis has as its major characteristic a progressive erosive and proliferative chronic polyarthritis with intermittent acute inflammatory episodes ("flares"). Cellular hypertrophy and hyperplasia, characteristics of malignant conversion induced by hypoxia [1, 2], occur within the synovial membrane. The ensuing "pannus" tissue protrudes into the joint space causing narrowing and infiltrates cartilage and bone where it is associated with erosive changes. This proliferating tissue exhibits poor neovascularity resulting in acidosis and low oxygen tension (pO_2). Synovial fluid volume is increased, particularly during "flares" and protein and lipid levels are elevated [3–6].

Although presenting with swollen joints and synovitis, RA is also manifested systemically and displays an immunopathology with circulating autoantibodies and cellular immune dysregulation [7–9]. T cell infiltration of the synovial membrane is

Inducible Enzymes in the Inflammatory Response, edited by D.A. Willoughby and A. Tomlinson
© 1999 Birkhäuser Verlag Basel/Switzerland

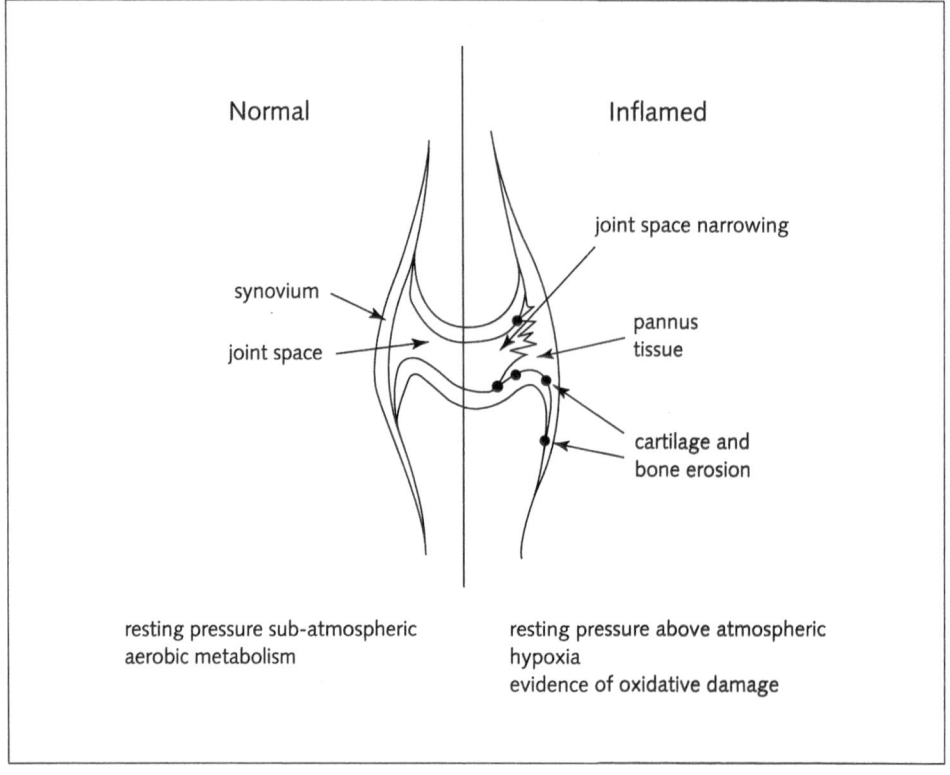

Figure 1
Schematic representation of the normal and inflamed diarthrodial synovial joint

apparent but the cytokine profile is predominantly macrophage- or fibroblast-derived (e.g. tumour necrosis factor α (TNFα), interleukin-1 (IL-1), IL-6, transforming growth factor β (TGFβ)) [10–12]. Recent studies have revealed that synovial tissue exhibits a Th1 type of helper T lymphocyte response, secreting interferon γ (IFNγ) but not IL-4 [13]. While the aetiology of RA remains an enigma, there is an association with certain HLA Class II antigens; these include DR4 and certain non-DR4 subtypes, encoded by the HLA-DRβ1 gene [14, 15].

Hypoxia, oxidative stress and rheumatoid arthritis

Although it is clear that arthritic joints are inherently hypoxic, evidenced by low glucose and elevated lactate levels, low pO_2 and hypercapnia, there is considerable evidence of oxidative modification to carbohydrates, proteins and cell membranes

[6, 16]. This paradox can be explained in part by considering the mechanics of joint function. The normal joint is an enclosed, weight-bearing and highly mobile environment. When inflamed, the volume of synovial fluid is increased but remains constrained within the joint capsule. Normal everyday movements (e.g. walking) create pressure fluxes within the joint; the resting pressure of a healthy knee joint is subatmospheric and falls slightly with exercise. Patients with arthritis display higher resting pressures (slightly above atmospheric) which rise further following movement to levels that exceed the capillary perfusion pressure; this has the effect of impeding blood flow through the synovium, creating a temporary ischaemia. On resting, reperfusion of the membrane occurs.

Using laser Doppler flowmetry we have confirmed that this mechanism operates in patients with RA, demonstrated by occlusion of the synovial capillary bed following quadriceps contraction and reperfusion on rest [17]. These cycles of ischaemia and re-oxygenation (redox cycling) lead to the release of reactive oxygen species (ROS), reduction products of molecular oxygen which can induce oxidative modification of biomolecules (hypoxia-reperfusion injury). Oxygen radicals are the only known metabolites of biological systems which can abstract methylene hydrogen atoms from polyunsaturated fatty acids (lipid peroxidation or rancidification), thereby damaging cell membranes. The most direct method for the detection of radical species is the technique of electron paramagnetic resonance spectroscopy (EPR), whereby radical intermediates are entrapped within specific chemicals (spin traps) and subjected to high magnetic fields: Spectra characteristic of different radicals are produced. Thus, we have demonstrated that excised synovium can generate radical species following redox (hypoxia/ normoxia) cycling [18, 19]. Consequently, episodes of oxidative stress will be superimposed on a chronically hypoxic environment, causing imbalance of cellular control systems.

Redox regulation, cellular stasis and inflammation

To endure, living cells must maintain and regulate intracellular pH levels and, to achieve this, an elegant series of iterating biochemical pathways have evolved. Ultimately cellular integrity (stasis) is governed by its reduction/ oxidation (redox) potential, the major intracellular redox regulator (buffer) being the tripeptide, glutathione. Thus, hypoxia leads to lowered pH and acidosis (reducing conditions), hyperoxia or oxidative stress induces reactive oxygen species (ROS) and reactive nitrogen intermediates (RNI) while normoxia represents physiological pH (stasis).

Evolutionary adaptation has resulted in the utilisation of oxygen (an electron acceptor) to provide energy in the form of ATP (oxidative phosphorylation), thereby regenerating NAD+ and FAD+. The molecular machinery, termed the electron transport chain, is localised to the chloroplast in plants and to the mitochondrial membrane in humans. Carrier molecules include nicotinamide nucleotides, heme proteins (cytochromes), flavoproteins and iron-sulphur proteins (e.g. xanthine oxidoreductase). Normal cells are maintained in a state of constant, if mild, oxidative stress due to leakage of electrons from the respiratory chain (estimated as approximately 1%), the constitutive release of nitric oxide (NO) by NO synthases and the liberation of ROS from oxidases and peroxidases [20].

Inflammation is a necessary response to tissue injury but, when uncontrolled, can lead to chronic inflammatory disease. One of the first responses to tissue injury is vasoconstriction (hypoxia) followed by activation of the clotting cascade. In wound healing, these anoxic conditions prevail, stimulating the release of angiogenic factors until vascularisation (reoxygenation) is completed. Gene activation is cell type specific, endothelial cells producing endothelin, macrophages releasing TNFα and hepatoma cells, erythropoietin [2]. Therefore, although continued oxygen starvation or hypoxia is potentially life threatening, temporary ischaemia is a necessary requirement for the normal inflammatory response since hypoxia provides a signal for triggering certain genes. This response to hypoxia can be manipulated to activate target genes [21].

Enzymes induced by hypoxia and oxidative stress

Nitric oxide synthase

Nitric oxide (nitrogen monoxide; NO), originally recognised as the endothelium-derived relaxing factor (EDRF), is an important messenger molecule for the regulation of vascular tone, platelet aggregation, neuronal signal transduction and immune-mediated responses [22–25]. The advent of NO as a regulatory molecule has transformed scientific opinion with respect to ROS. Once considered purely damaging or too transient to be plausible as a signalling system, it is now recognised that both ROS and RNI have two major functions: cytotoxicity and cell-cell communication.

NO is synthesised from the amino acid L-arginine by nitric oxide synthases (NOS) and therefore arginine analogues can be used as inhibitors of NOS. The inducible enzyme (iNOS), located in macrophages, is induced by endotoxin and pro-inflammatory cytokines (TNFα, IFNγ and IL-1β) in a NADPH-dependent and calcium-independent process and down-regulated by IL-8, TGFβ and platelet derived growth factor (PDGF). Constitutive isoforms of NOS (cNOS) occur on endothelial (eNOS) and neuronal (nNOS) cells and require calcium/ calmodulin for activation. The mechanism of NO release varies between the isoforms; release of NO from cNOS is immediate and short-lived whereas expression of the inducible isoform is delayed (4–6 hours) but the magnitude and duration of NO release is enhanced [26, 27]. NOS are induced during hypoxia/ reperfusion events (discussed below).

Cyclooxygenase

Cyclooxygenase (prostaglandin H synthase) mediates the formation of prosta-glandins (PG) from arachidonic acid (AA) [28]. Again, constitutive (COX-1) and inducible (COX-2) isoforms exist. COX-1 is present in the stomach, gut and kidney while COX-2 is induced by pro-inflammatory cytokines and hypoxia and expressed in various cell types (eg fibroblasts and macrophages) at inflammatory sites; COX-2 but not COX-1 synthesis is markedly suppressed by corticosteroids [29]. Hypox-ia-induced COX-2 expression in endothelial cells [30] and IL-1-induced COX-2 expression in synoviocytes [31] is dependent on NFκB binding. Selective COX-2 inhibition is anti-inflammatory and inhibits nociception [32].

Interestingly, both COX-1 and COX-2 are activated by NO and prostaglandin release is inhibited by treatment with iNOS inhibitors. However, selective inhibition of COX-2 has little effect on NO production. Salvemini [26] has suggested that this may explain why, in arthritis, a compound such as indomethacin, by blocking PG but not NO production, alleviates the inflammatory component but has little effect on the disease course.

Heme oxygenase

Heme oxygenase is the rate-limiting step in the conversion of heme to bilirubin. Like NOS and COX, inducible (HO-1) and constitutive (HO-2) isoforms exist and lack of the enzyme causes hyperbilirubinaemia, particularly in premature newborn infants [33]. HO-2 is located in mitochondria while HO-1 is a microsomal enzyme now recognised as the 32 kDa stress or heat shock protein (hsp32) [33–36]. Although described as a marker of oxidative stress, hsp32 can be induced by numer-ous stressors including elevated temperature, cytokines, UVA radiation, hydrogen peroxide and heavy metal ions as well as its substrate heme; the enzyme activity is inhibited by tin and zinc protoporphyrins and upregulated by ferriprotoporphyrin (hemin) [33, 35]. Glutathione depletion is a powerful inducer of HO-1 and therefore the enzyme is more correctly termed a marker of redox imbalance [36, 37]. Inter-estingly, the anti-rheumatic compound disodium aurothiomalate ("gold") is anoth-er potent inducer of HO-1 [38]. Lipid peroxidation products also induce HO-1, [39] whilst arachidonate [40] and prostaglandins [41, 42] can modulate hsp expression.

Heme oxygenase converts heme to biliverdin, releasing Fe^{2+} and carbon mono-xide (CO) [33, 43]. In health, Fe^{2+} is sequestered into ferritin while CO exhibits reg-ulatory properties similar to NO, both acting through the generation of cGMP [43, 44]. Moreover, NO activates HO-1 [45] and CO is an inhibitor of iNOS activity [46]; therefore the two enzymes may have synergistic effects. In endothelial cells *in vitro*, both activation of iNOS and addition of NO donors augmented HO activa-tion but addition of iNOS inhibitors had no effect on HO activity [45].

Xanthine oxidoreductase

The enzyme xanthine oxidoreductase (XOR) converts hypoxanthine and xanthine to uric acid. Two forms of the enzyme exist and, in fact, uric acid production is dependent on the dehydrogenase form of the enzyme (XaD) which donates electrons to NAD^+, thereby generating NADH. Under hypoxic conditions, however, the enzyme is rapidly converted to its oxidase form (XaO) which donates electrons to molecular oxygen forming superoxide and other ROS. The damage induced by this redox cycling has been termed hypoxia/ reperfusion injury [47, 48].

At the cellular level, oxidative phosphorylation ceases, ATP levels fall and energy generation becomes dependent on anaerobic glycolysis with the concomitant depletion of glucose in favour of lactate. Intracellular calcium levels rise while ATP continues to decline, leading to the accumulation of adenosine and its breakdown products xanthine and hypoxanthine, substrates for XOR.

Conversion of XaD to XaO is induced by a calcium-dependent protease which is activated by hypoxia; however, enzymic conversion may be effected by oxidants themselves following the oxidation of certain thiol groups on XaD and also by the monokine, TNFα. TNFα not only causes XaD to XaO conversion but may also transcriptionally upregulate mRNA expression of the enzyme in endothelial cells, the immediate target of reperfusion injury [49].

XOR exists as a dimer, each sub-unit containing one molybdopterin, two non-identical iron-sulphur centres and FAD. Because of this molecular structure, the enzyme has broad substrate specificity and exhibits numerous activities. The human enzyme favours NADH, rather than xanthine, as a substrate for superoxide generation; however, at low pO_2, the enzyme has nitrate and nitrite reductase capacity, utilising either xanthine or NADH as electron donors to generate NO [50].

Hypoxia, oxidative stress and gene expression

Gene transcription

The basic premise of gene expression is that the nucleotide sequence (DNA coding region) of a gene dictates the form of the expressed protein, for example, an enzyme. Full gene expression involves activation, transcription (RNA synthesis) and translation (protein synthesis) but the major control point lies at the level of transcription. The gene promotor region, lying immediately upstream of the gene coding region contains short DNA sequences (elements) which are control points for transcription. Examples include glucocorticoid, metal and cAMP responsive elements (GRE, MRE, CRE respectively) as well as binding sites for specific transcription factors (TFs). More recently, antioxidant and hypoxia responsive elements (ARE and HRE) have been described as well as a specific hypoxia-inducible transcription factor

(HIF). These sites may occur singly or in multiples within individual genes. Therefore, the cellular environment can control gene expression (e.g. high cAMP levels activating a specific gene pool through the CRE) and aberrant signalling can result in the induction of unwanted proteins (e.g. enzymes). However, for full activation, simultaneous binding of several sites may be required.

With this in mind, it is easy to perceive that inflammation will effect gene expression leading to the induction of specific proteins (e.g. enzymes). Thus, hypoxia or antioxidant loading can lead to the release of metal ions from storage proteins (which can bind to MRE) or increased levels of cAMP (which may activate CRE dependent genes), while oxidative stress or hyperoxia can initiate induction of the immediate early genes ("oncogenes": *c-fos*, *c-myc*, *c-jun*), inducing proliferation. Moreover, it is now clear that more subtle regulation of certain transcription factors may be mediated by radical species [51, 52].

Redox-controlled transcription factors

The oxidant response factor – NFκB
Nuclear factor kappa B (NFκB) was the first TF shown to respond directly to oxidative stress [53]. NFκB is a heterodimer which is held in an inactive state in the cytoplasm by binding of an inhibitory factor, IκB. Upon activation, NFκB and IκB dissociate and active NFκB translocates to the nucleus; many of its target genes encode proteins central to inflammation and immune recognition including cytokines (IL2 and TNFα), cytokine receptors (IL2R), MHC Class 1 antigens, adhesion molecules (VCAM-1) and, of course, the immunoglobulin kappa chain from which its name derives. As well as controlling cytokine gene expression, NFκB is itself activated by the inflammatory cytokines IL1 and TNFα [54]. Many activators of NFκB also induce oxidative stress indicating that, at low concentrations, ROS may fulfil a common second messenger role. Interestingly, it has been shown that the DNA-binding activity of NFκB is controlled by at least two cellular redox mechanisms, thioredoxin/ thioredoxin reductase and apurinic/ apyrimidinic endonuclease (Redox factor-1; Ref-1) [55]. Thus, thioredoxin is a potent co-stimulator of cytokine expression and thioredoxin reductase is another inducible enzyme, upregulated by oxidative stress. Thioredoxin attenuates hypoxia/ reperfusion injury in model systems [56, 57].

The antioxidant response factor – AP-1
A second redox-sensitive transcription factor is the activator protein-1 (AP-1) [52]. This exists in two forms as either a homodimer of c-jun or a heterodimer of c-jun with c-fos, both immediate early gene products. Again inducers of oxidative stress including uv light, gamma irradiation or IL1 activate AP-1. However, to achieve

gene activation, AP-1 must be in dimeric form and it has been shown that H_2O_2 induces strong expression of c-fos and/or c-jun with little formation of AP-1 dimers. Conversely, *anti*oxidants induce c-fos and c-jun with significant AP-1 dimer formation and substantial DNA-binding activity. This antioxidant-induced binding is dependent on protein synthesis and was shown to require a third redox-sensitive transcription factor, SRF/TCF, at the level of c-fos [51]. The redox-regulated serum response factor (SRF) and the ternary complex factor (TCF) bind to adjacent elements on the c-fos promotor although TCF activation may require phosphorylation [51]. Interestingly, two further enzyme systems are activated following oxidative stress, the mitogen-activated protein (MAP) kinase cascade (phosphorylating) and a MAP kinase phosphatase [58] (dephosphorylating). Moreover, like NFκB, the DNA binding activity of AP-1 has been linked to thioredoxin-mediated redox modulation of the DNA repair enzyme, Ref-1 [59].

Transcription factors and inducible enzymes

COX-2 activation is dependent on NFκB binding [30, 31], iNOS can be activated through a HRE [60] and the HO-1 promotor contains numerous TF elements including AP-1 and NFκB-like sites as well as MRE and ARE [61–63]. Moreover, specific heat shock transcription factors (HSF) have been described: HSF-1 controls the response to heat shock, oxidative stress and heavy metals while HSF-2 is required for hsp induction by hemin and during embryonic development [64]. Although four HSF are now recognised, not all of them are well characterised. As outlined above, XOR shows post translational modification which modulates its activity. We have immunohistological evidence of enhanced active NFκB expression in arthritic synovia [65].

Inducible enzymes and rheumatoid arthritis

Histological localisation

In an attempt to determine the importance of inducible proteins in inflammatory disease, the first approach is often indirect immunohistochemical localisation using antibodies. Clearly, this type of analysis is highly dependent on the specificity of the antibody and therefore, more recently, gene probes have been employed. Nevertheless, the implication that presence indicates activity is not always relevant and this type of data may be overinterpreted. With this caveat in mind, all four of the inducible enzymes listed above have been demonstrated in tissues from rheumatoid joints although they are not specific to RA since they are found in other inflammatory diseases.

The enzyme iNOS was shown to be strongly expressed in the synovial lining layer, subsynovium, vascular smooth muscle and chondrocytes from patients with RA and to a lesser extent in osteoarthritis: iNOS was not upregulated in normal joint tissues. The major source was identified as CD68+ macrophages in the synovial lining layer [66]. COX-2 was shown to be expressed in infiltrating mononuclear cells, endothelial cells of blood vessels and subsynovial fibroblastic cells but the lining layer stained weakly: COX-2 transcripts were detectable in the mRNA of synovial explants [29, 31]. Our group has demonstrated expression of both xanthine oxidoreductase (XOR) and HO-1 in synovial tissue. XOR was localised to endothelial cells [67, 68] while HO-1 was demonstrable not only in vessel walls but in intimal cells and stromal fibroblasts, although the pattern of staining showed individual variation. Synovial explants contained high levels of HO-1 mRNA [69].

Implications for pathogenesis and therapy

Evidence of interactions between products of inducible enzymes is accumulating rapidly but their precise roles in inflammation remain speculative. The xanthine oxidase system, present both on the plasma membrane and within the cytoplasm appears to function in tandem with NOS. NO and superoxide can interact rapidly, generating the non-radical species, peroxynitrite ($ONOO^-$), which is not only toxic in itself but renders numerous highly toxic products including nitrogen dioxide gas ($NO_2\cdot$), the highly reactive hydroxyl radical ($\cdot OH$) and the nitronium ion (NO_2^+) [70]. Aromatic amino acids are particularly prone to the nitrating effects of these species and the formation of 3-nitrotyrosine is now recognised as a biomarker, not only of peroxynitrite generation, but of nitrating species in general [71]. This oxidant may prove to be of particular importance in inflammatory disease since the two radical generating systems may associate following adhesion of neutrophils to endothelium. Indeed, we have immunohistological evidence of peroxynitrite formation in the rheumatoid synovium [72].

Figure 2 is a schematic representing the effects of hypoxia reperfusion events on the regulation of inducible enzymes. The figure represents only the direct effects of redox cycling: cytokines also upregulate inducible enzymes. TNFα, a major pro-inflammatory cytokine present in inflamed joints, induces both iNOS and HO-1 and can mediate the conversion of XaD to its oxidase form. In clinical trials, blockade of TNFα with monoclonal antibodies was beneficial in short term trials [73, 74] but did not halt disease progression; moreover, the beneficial effects waned with continued treatment [73].

In patients with RA, both synovium and cartilage produce NO [75] and bone erosions have been linked with cytokine-induced NO release [76]; we were the first group to report elevated levels of nitrite in serum and synovial fluids, suggestive of enhanced iNOS activity [77]. Of importance here is that at low pO_2, XOR

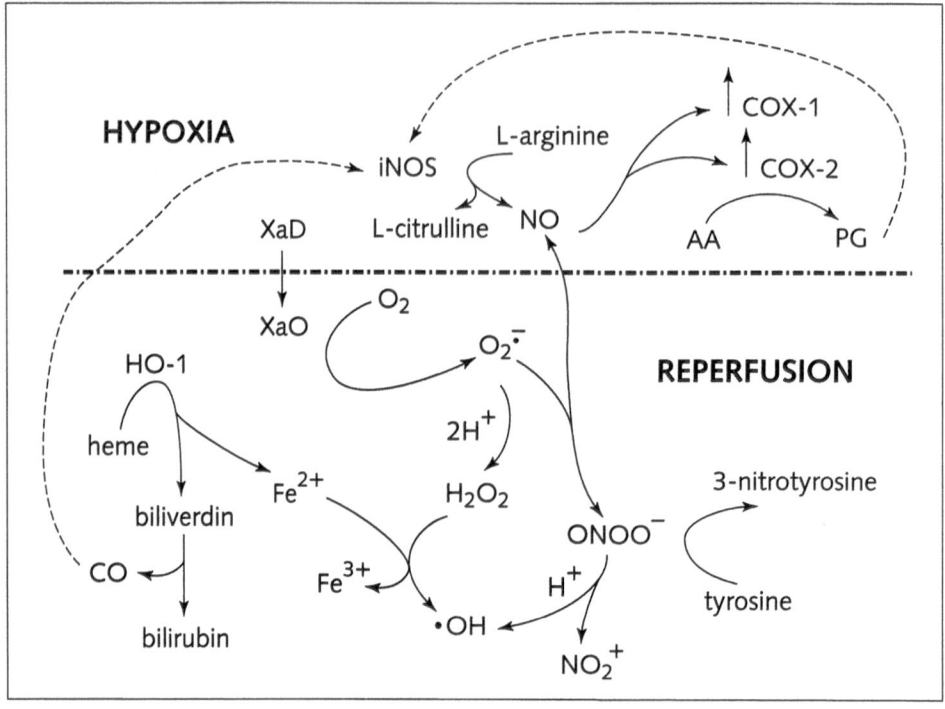

Figure 2

Hypoxia-reperfusion events (redox cycling) contribute to the release of inducible enzymes and the persistence of rheumatoid synovitis.

During hypoxia (ischaemia), xanthine dehydrogenase (XaD), inducible nitric oxide synthase (iNOS) and cyclo-oxygenase-2 (COX-2) are all upregulated. Reperfusion mediates conversion of xanthine dehydrogenase (XaD) to xanthine oxidase (XaO) and induces heme oxygenase-1 (HO-1). The reactive intermediates generated may associate to form further toxic products. Both carbon monoxide (CO) and prostaglandins (PG) are known to cause feedback inhibition (----) of iNOS while PG can upregulate HO-1 activity leading to further release of CO. In the mobile inflamed joint, constant redox cycling may impair the balance of regulation between these enzymes.

has nitrite reducing activity, utilising either xanthine or NADH as electron donors [50].

Unquestionably, the complex nature of interactions between these enzymes provides scope for novel therapies. COX-2 inhibitors are already in use in patients with RA [32] and iNOS inhibitors have shown efficacy in disease models [25]. HO-1 antagonists have been used to treat hyperbilirubinaemia of newborn infants but side effects occur [78]; these antagonists are being investigated in model systems of inflammation, including arthritis [79]. Allopurinol, a XOR inhibitor, is useful in the treatment of gout but not in RA, possibly because the nitrite reductase activity of XOR is not inhibitable with this pharmacological agent [50]. Clearly, further analysis of these iterating systems is necessary before their therapeutic usefulness is known.

Acknowledgements
Professor Blake is The Glaxo-Wellcome Professor of Bone and Joint Medicine at The University of Bath and Dr Winrow is funded by The Arthritis Research Campagin, Chesterfield, UK [Grant No B0568]. The authors wish to thank the Arthritis Research Campaign for continuing support [Grant Nos B0511, B0583].

References

1   Gay S, Gay RE (1989) Cellular basis and oncogene expression of rheumatoid joint destruction. *Rheumatol Int* 9: 105–113
2   Anderson GR, Stoler DL (1993) Anoxia, wound-healing, VL30 elements, and the molecular basis of malignant conversion. *Bioessays* 15: 265–272
3   Gardner DL (ed) (1972) *The Pathology of Rheumatoid Arthritis*. Edward Arnold Publishers, London
4   Fassbender HG (1983) The potential agressiveness of synovial tissue in rheumatoid arthritis. *J Pathol* 139: 399–406
5   Henderson B, Pettipher ER (1985) The synovial lining cell: Biology and immunopathology. *Semin Arthritis Rheum* 15: 1–32
6   Stevens CR, Williams RB, Farrell AJ, Blake DR (1991) Hypoxia and inflammatory synovitis: Observations and speculation. *Ann Rheum Dis* 50: 124–132
7   Panayi GS, Lanchbury JS, Kingsley GH (1992) The importance of the T cell in initiating and maintaining the chronic synovitis of rheumatoid arthritis. *Arthritis Rheum* 35: 729–735
8   Sewell KL, Trentham DE (1993) Pathogenesis of rheumatoid arthritis. *Lancet* 341: 283–290
9   Edwards JCW, Cambridge J (1995) Is rheumatoid arthritis a failure of B cell death in synovium? *Ann Rheum Dis* 54: 696–700
10 Feldmann M, Brennan FM, Maini RM (1996) Rheumatoid arthritis [review]. *Cell* 85: 307–310
11 Arend WP, Dayer J-M (1990) Cytokines and cytokine inhibitors or antagonists in rheumatoid arthritis. *Arthritis Rheum* 33: 305–315
12 Arend WP (1997) The pathophysiology and treatment of rheumatoid arthritis [editorial]. *Arthritis Rheum* 40: 595–597

13 Simon AK, Seipelt E, Seiper J (1994) Divergent T cell patterns in inflammatory arthritis. *Proc Natl Acad Sci USA* 91: 8562–8566

14 Gregersen PK, Silver J, Winchester RJ (1987) The shared epitope hypothesis: An approach to understanding the molecular genetics of susceptibility to rheumatoid arthritis. *Arthritis Rheum* 30: 1205–1213

15 Ollier W, Thomson W (1992) Population genetics of rheumatoid arthritis. *Rheum Dis Clin N Am* 18: 761–785.

16 Winrow VR, Winyard PG, Morris CJ, Blake DR (1993) Free radicals in inflammation: Second messengers and mediators of tissue destruction. *Br Med Bull* 49: 506–522

17 Blake DR, Merry P, Unsworth J, Kidd BL, Outhwaite JM, Ballard R, Morris CJ, Gray L, Lunec J (1989) Hypoxic-reperfusion injury in the inflamed human joint. *Lancet* i: 289–293

18 Allen RE, Blake DR, Nazhat NB, Jones P (1989) Superoxide radical generation in inflamed human synovium after hypoxia. *Lancet* ii: 282–283

19 Singh D, Nazhat NB, Fairburn K, Sahinoglu T, Blake DR, Jones P (1995) Electron spin resonance spectroscopic demonstration of the generation of reactive oxygen species by diseased human synovial tissue following ex vivo hypoxia-reoxygenation. *Ann Rheum Dis* 54: 94–99

20 Halliwell B (1994) Free radicals and antioxidants: A personal view. *Nutr Rev* 52: 253–265

21 Dachs GU, Patterson AV, Firth JD, Ratcliffe PJ, Townsend KMS, Stratford IJ, Harris AL (1997) Targeting gene expression to hypoxic tumour cells. *Nature Med* 3: 515–520

22 Palmer RMJ, Ferrige AG, Moncada S (1987) Nitric oxide release accounts for the biological activity of endothelium-derived relaxing factor. *Nature* 327: 524–526.

23 Furchgott RF (1996) The discovery of endothelium-derived relaxing factor and its importance in the identification of nitric oxide. *JAMA* 276: 1186–1188

24 Moncada S, Higgs EA (1995) Molecular mechanisms and therapeutic strategies related to nitric oxide. *FASEB J* 9: 1319–1330

25 Evans CH (1995) Nitric oxide: What role does it play in inflammation and tissue destruction? *Agents Actions* (suppl) 47: 107–116

26 Salvemini D (1997) Regulation of cyclooxygenase enzymes by nitric oxide. *Cell Mol Life Sci* 53: 576–582

27 Ralston SH (1997) Nitric oxide and bone: What a gas! *Br J Rheumatol* 36: 831–838

28 DeWitt DL (1991) Prostaglandin endoperoxide synthase: Regulation of enzyme expression. *Biochim Biophys Acta* 1083: 121–134

29 Crofford LJ, Wilder RL, Ristimaki AP, Remners EF, Epps HR, Hla T (1994) Cyclooxygenase-1 and -2 expression in rheumatoid synovial tissues: Effects of interleukin-1β, phorbol ester, and corticosteroids. *J Clin Invest* 93: 1095–1101

30 Schmedtje JF, Ji Y-S, Liu W-L, DuBois RN, Runge MS (1997) Hypoxia induces cyclooxygenase-2 via the NF-κB p65 transcription factor in human vascular endothelial cells. *J Biol Chem* 272: 601–608.

31 Crofford LJ, Tan B, McCarthy CJ, Hla T (1997) Involvement of nuclear factor κB in the

regulation of cyclooxygenase-2 expression by interleukin-1 in rheumatoid synoviocytes. *Arthritis Rheum* 40: 226–236

32   Emery P (1996) Clinical implications of selective cyclooxygenase-2 inhibition. *Scand J Rheumatol* 25 (suppl): 23–28

33   Maines MD (1988) Heme oxygenase: function, multiplicity, regulatory mechanisms, and clinical applications. *FASEB J* 2: 2557–2568

34   Keyse SM, Tyrrell RM (1989) Heme oxygenase is the major 32-KDa stress protein induced in human skin fibroblasts by UVA radiation, hydrogen peroxide, and sodium arsenite. *Proc Natl Acad Sci USA* 86: 99–103

35   Donati YR, Slosman DO, Polla BS (1990) Oxidative injury and the heat shock response. *Biochem Pharmacol* 40: 2571–2577

36   Applegate LA, Lüscher P, Tyrrell RM (1991) Induction of heme oxygenase: A general response to oxidant stress in cultured mammalian cells. *Cancer Res* 51: 974–978

37   Lautier D, Lüscher P, Tyrrell RM (1992) Endogenous glutathione levels modulate both constitutive and UVA radiation/hydrogen peroxide inducible expression of the human heme oxygenase gene. *Carcinogenesis* 13: 227–232

38   Caltabiano MM, Poste G, Grieg RG (1986) Induction of the 32-kD human stress protein by auranofin and related triethylphosphine gold analogs. *Biochem Pharmacol* 37: 4089–4093

39   Basu-Modak S, Lüscher P, Tyrrell RM (1996) Lipid metabolite involvement in the activation of the human heme oxygenase-1 gene. *Free Rad Biol Med* 20: 887–897

40   Jurivich DA, Sistonen L, Sarge KD, Morimoto RI (1994) Arachidonate is a potent modulator of human heat shock gene transcription. *Proc Natl Acad Sci USA* 91: 2280–2284

41   Koizumi T, Negishi M, Ichikawa A (1992) Induction of heme oxygenase by delta 12-prostaglandin J2 in porcine aortic endothelial cells. *Prostaglandins* 43: 121–131

42   Rossi A, Santoro MG (1995) Induction by prostaglandin A1 of haem oxygenase in myoblastic cells: An effect independent of expression of the 70kDa heat shock protein. *Biochem J* 308: 455–463

43   Maines MD (1996) Carbon monoxide and nitric oxide homology: Differential modulation of heme oxygenase in brain and detection of protein and activity. *Methods Enzymol* 268: 473–488

44   Ignarro LJ (1992) Haem-dependent activation of cytosolic guanylate cyclase by nitric oxide: A widespread signal transduction mechanism. *Biochem Soc Trans* 20: 465–469

45   Motterlini R, Foresti R, Italglietta M, Winslow RM (1996) NO-mediated activation of heme oxygenase: Endogenous cytoprotection against oxidative stress to endothelium. *Am J Physiol* 270: H107–114

46   White KA, Marletta MA (1992) Nitric oxide synthase is a cytochrome P-450 type hemoprotein. *Biochemistry* 31: 6627–6631

47   Della Corte E, Stirpe F (1972) The regulation of rat liver xanthine oxidase: Involvement of thiol groups in the conversion of the enzyme activity from the dehydrogenase (type D) to the oxidase (type O) and purification of the enzyme. *Biochem J* 126: 739–745

48   Granger DN, Rutili G, McCord JM (1981) Superoxide radicals in feline intestinal ischaemia. *Gastroenterology* 81: 22–29

49   Friedl HP, Till GO, Ryan US, Ward PA (1989) Mediator-induced activation of xanthine oxidase in endothelial cells. *FASEB J* 3: 2512–2518

50   Blake DR, Stevens CR, Sahinoglu T, Ellis G, Gaffney K, Edmonds S, Benboubetra M, Harrison R, Jawed S, Kanczler J et al (1997) Xanthine oxidase: four roles for the enzyme in rheumatoid pathology. *Biochem Soc Trans* 25: 812–816

51   Meyer M, Schreck R, Baeuerle PA (1993) H_2O_2 and antioxidants have opposite effects on activation of NF-κB and AP-1 in intact cells: AP-1 as secondary antioxidant-responsive factor. *EMBO J* 12: 2005–2015.

52   Pahl HL, Baeuerle PA (1994) Oxygen and the control of gene expression. *Bioessays* 16: 497–502

53   Schreck R, Albermann K, Baeuerle PA (1992) Nuclear factor kappa B: an oxidative stress-responsive transcription factor of eukaryotic cells. *Free Rad Res Comms* 17: 221–237

54   Baeuerle PA, Baltimore D (1991) The physiology of the NF-κB transcription factor. In: Cohen P, Foulkes JG (eds) *The hormonal control regulation of gene transcription.* Elsevier Science Publishers, Amsterdam, 423–446

55   Mitomo K, Nakayama K, Fujimoto K, Sun X, Seki S, Yamamoto K (1994) Two different cellular redox systems regulate the DNA-binding activity of the p50 subunit of NF-κB *in vitro. Gene* 145: 197–203

56   Fukuse T, Hirata T, Yokomise H, Hasegawa S, Inui K, Mitsui A, Hirakawa T, Hitomi S, Yodoi J, Wada H (1995) Attenuation of ischaemia reperfusion injury by human thioredoxin. *Thorax* 50: 387–391

57   Aota M, Matsuda K, Isowa N, Wada H, Yodoi J, Ban T (1996) Protection against reperfusion-induced arrhythmias by human thioredoxin. *J Cardiovascular Pharmacol* 27: 727–732

58   Keyse SM, Emslie EA (1992) Oxidative stress and heat shock induce a human gene encoding a protein-tyrosine phosphatase. *Nature* 359: 644–647

59   Hirota K, Matsui M, Iwata S, Nishiyama A, Mori K, Yodoi J (1997) AP-1 transcriptional activity is regulated by a direct association between thioredoxin and Ref-1. *Proc Natl Acad Sci USA* 94: 3633–3638

60   Melillo G, Musso T, Sica A, Taylor LS, Cox GW, Varesio L (1995) A hypoxia-responsive element mediates a novel pathway of activation of the inducible nitric oxide synthase promotor. *J Exp Med* 182: 1683–1693

61   Lavrovsky Y, Schwartzman ML, Levere RD, Kappas A, Abraham NG (1994) Identification of binding sites for transcription factors NFκB and AP-2 in the promoter region of the human heme oxygenase 1 gene. *Proc Natl Acad Sci USA* 91: 5987–5991

62   Takeda K, Ishizawa S, Sato M, Yoshida T, Shibahara S (1994) Identification of a *cis*-acting element that is responsible for cadmium-mediated induction of the human heme oxygenase gene. *J Biol Chem* 269: 22858–22867

63   Prestera T, Talalay P, Alam J, Ahn YI, Lee PJ, Choi AMK (1995) Parallel induction of

heme oxygenase-1 and chemoprotective phase 2 enzymes by electrophiles and antioxidants: Regulation by upstream antioxidant-responsive elements (ARE). *Mol Med* 1: 827–837

64   Morimoto RI (1993) Cells in stress: Transcriptional activation of heat shock genes. *Science* 259: 1409–1410

65   Marok R, Winyard PG, Coumbe A, Kus ML, Gaffney K, Blades S, Mapp PI, Morris CJ, Blake DR, Kaltschmidt C, Baeuerle PA (1996) Activation of the transcription factor nuclear factor-kappa B in human inflamed synovial tissue. *Arthritis Rheum* 39: 583–591

66   Grabowski PS, Wright PK, Van'T Hof RJ, Helfrich MH, Ohshima H, Ralston SH (1997) Immunolocalization of inducible nitric oxide synthase in synovium and cartilage in rheumatoid arthritis and osteoarthritis. *Br J Rheum* 36: 651–655

67   Allen RE, Outhwaite JM, Morris CJ, Blake DR (1987) Xanthine oxido-reductase is present in human synovium. *Ann Rheum Dis* 46: 834–845

68   Stevens CR, Benboubetra M, Harrison R, Sahinoglu T, Smith EC, Blake DR (1991) Localisation of xanthine oxidase to synovial endothelium. Ann Rheum Dis 50: 760–762

69   Winrow VR, Mapp PI, Watson A, Gaffney K, Blake DR (1995) 32kDa stress protein expression in arthritic synovia. *Br J Rheumatol* 34 (abstr suppl 1): 47

70   Beckman JS, Crow JP (1993) Pathological implications of nitric oxide, superoxide and peroxynitrite formation. *Biochem Soc Trans* 21: 330–334

71   Halliwell B (1997) What nitrates tyrosine? Is nitrotyrosine specific as a biomarker of peroxynitrite formation *in vivo*. *FEBS Lett* 411: 157–160.

72   Blake DR, Blades S, Coumbe A, Mapp PI (1996) Nitration of tyrosine in the inflamed human synovium: Evidence for the generation of peroxynitrite *in vivo*. *Br J Rheumatol* 35 (abstr suppl 1): 70

73   Maini RN, Elliott MJ, Brennan FM, Feldmann M (1995) Beneficial effects of tumour necrosis factor-alpha (TNF-α) blockade in rheumatoid arthritis (RA). *Clin Exp Immunol* 101: 207–212

74   Rankin ECC, Choy EHS, Kassimos D, Kingsley GH, Sopwith AM, Isenberg DA, Panayi GS (1995) The therapeutic effects of an engineered human anti-tumour necrosis factor alpha antibody (CDP571) in rheumatoid arthritis. *Br J Rheumatol* 34: 334–342

75   Sakurai H, Kohsaka H, Liu M-F, Higashiyama H, Hirata Y, Kanno K, Saito I, Miyasaka N (1995) Nitric oxide production and inducible nitric oxide synthase expression in inflammatory arthritides. *J Clin Invest* 96: 2357–2363

76   Ralston SH (1997) Nitric oxide and bone: What a gas! *Br J Rheumatol* 36: 831–838

77   Farrell AJ, Blake DR, Palmer RMJ, Moncada S (1992) Increased concentrations of nitrite in synovial fluid and serum samples suggest increased nitric oxide synthesis in rheumatic diseases. *Ann Rheum Dis* 51: 1219–1222

78   Kappas A, Drummond GS, Simionatto CS, Anderson KE (1984) Control of haem oxygenase and plasma levels of bilirubin by a synthetic heme analogue, tin-protoporphyrin. *Hepatology* 4: 336–341

79   Willis D, Moore AR, Frederick R, Willoughby DA (1996) Heme oxygenase: A novel target for the modulation of the inflammatory response. *Nature Med* 2: 87–90.

iNOS and COX-2 in atherosclerosis

Lee D.K. Buttery and Julia M. Polak

Department of Histochemistry, Imperial College School of Medicine, Hammersmith Campus, Du Cane Road, London W12 0NN, UK

Introduction

Atherosclerosis is a disease of the blood vessels and its associated clinical events are a significant cause of mortality and morbidity in the Western world today [1, 2]. The disease is regulated by a large number of different factors which partly explains why the aetiology and pathogenesis of atherosclerosis remain incompletely defined. Since the discovery of prostacyclin in the 1970s [3] and nitric oxide in the 1980s [4] considerable evidence has been amassed to indicate that these physiologic and pathophysiologic mediators contribute significantly to the regulation of atherosclerosis. The aim of this short review is to discuss the actions of these mediators in atherosclerosis, focusing on the enzymes regulating their synthesis, in particular their inducible isoforms.

Atherosclerosis

Increasing evidence associates chronic inflammation of the blood vessel wall with the development of atherosclerosis. The disease develops in the intima of medium to large sized arteries, principally affecting the abdominal aorta, coronary, carotid, iliac and popliteal arteries. Specific locations in these blood vessels are affected representing sites of turbulent flow or branching. The disease appears to begin in infancy, as suggested by the presence in the blood vessels of newborn babies of "fatty streaks", believed to be one of the earliest lesions associated with the development of atherosclerosis. Thereafter the disease appears to progress slowly throughout adulthood. In general, atherosclerosis remains clinically silent until the fourth or fifth decade of life and onwards, when the function and morphology of the vessel become so severely disrupted by the disease that the supply of blood to the organ/tissue that the vessel serves is attenuated. The pathophysiological manifestations of such disturbances are conditions such as angina pectoris, stroke or gangrene of the extremities [1, 2].

Inducible Enzymes in the Inflammatory Response, edited by D.A. Willoughby and A. Tomlinson
© 1999 Birkhäuser Verlag Basel/Switzerland

Atherogenesis is a complex process involving multiple cell types including the endothelium, monocytes/macrophages, lymphocytes, platelets and smooth muscle cells. In conjunction with blood-borne lipids and a wide-spectrum of cytokines and growth factors derived from these various cell types, atherosclerosis is marked by the development of distinctive morphological lesions progressing from the early lesion of the fatty streak through to the advanced plaque. Genetics, gender and lifestyle, in particular diet and smoking habit are also highly significant factors in the pathogenesis of this disease [1, 2]. Central to the initiation of atherogenesis are sites of localized endothelial dysfunction which act as foci for the stimulation and potentiation of a chronic inflammatory response within the intima of the vessel [1, 2]. It is also becoming apparent that the macrophage and oxidized lipids, in particular low density lipoprotein (LDL) entrapped in the vessel wall have a pivotal role in orchestrating this inflammatory response [1, 2, 5, 6].

Some of the features of atherosclerosis can be induced in animal models by exposing them to certain risk factors such as high cholesterol diet which may predispose to, or exacerbate the development of the disease [1, 2]. While such studies have proved useful in gaining a better understanding of the mechanism of atherogenesis, the aetiology of the disease still remains equivocal. Numerous hypotheses have been elaborated to explain the disease process. Of these, the "response to injury" hypothesis formulated by Ross and Glomset [7] and subsequently modified [1, 8] is perhaps the most widely accepted hypothesis at the present time. In essence, it states that atherogenesis results from a localized inflammatory response, which in the first instance is protective, but with time and continuing insult this response becomes excessive and ultimately forms the basis of the disease process itself.

NO synthase(s)

Nitric oxide (NO) is enzymatically produced by the oxidation of one or more of the terminal nitrogen atoms of the amino acid L-arginine in a reaction which utilizes molecular oxygen and also yields L-citrulline. This reaction is enantiometrically specific as D-arginine is not a substrate and it is also inhibited by arginine analogues such as L-NMMA and L-NAME. NO is involved in a variety of physiological and pathophysiological process including vasodilatation, neurotransmission, the killing of various pathogens and in the pathogenesis of inflammatory disease. These broad ranging actions of NO are determined largely by the site of NO synthesis, by how much is generated and by the nature of the environment into which NO is released. The reactivity of NO is influenced in particular by the presence of reactive oxygen intermediates and the activity of anti-oxidant defence systems [9, 10].

A family of three related enzymes – the NO synthases, regulates the synthesis of NO. Two enzymes are constitutively expressed; one present mainly in neural tissues (nNOS) and the other mainly in endothelial cells (eNOS). The constitutive enzymes are in general associated with the low level generation of NO which has a tonic, physiologic function [9]. The third isoform, the inducible NOS (iNOS), is not usually present in normal resting cells but its expression can be induced in a vast number of different cell types in response to infection, inflammation or traumatic tissue injury. Once activated, iNOS is capable of sustained generation of NO over many hours, which can in some circumstances lead to localized production of high levels of NO [9, 10]. The almost universal ability of the body's cells and tissues to generate NO via this high-output pathway has prompted the suggestion that the iNOS pathway represents an ancient (evolutionary) and relatively rapid immune effector system [11]. Stimulated overproduction however, can contribute to the pathology of many inflammatory diseases including sepsis, asthma, rheumatoid arthritis, dilated cardiomyopathy [9, 10] and, as will be discussed in this chapter, atherosclerosis. Concurrent generation of NO and reactive oxygen intermediates, in particular superoxide anion, at sites of inflammation favours the formation of the NO-derived species peroxynitrite [12]. Peroxynitrite is a powerful oxidant species, which may further exacerbate the tissue damaging potential of the high-output iNOS pathway [12, 13]. As a consequence of these potential deleterious effects there is currently much interest in controlling the expression and activity of iNOS.

NO synthase(s) and atherosclerosis

The generation of NO by the endothelium is fundamental to normal vascular function and its actions are considered to be anti-thrombotic and anti-atherogenic. These activities include inhibition of platelet aggregation [14], modulation of both platelet and monocyte-endothelial cell interactions [15, 16], inhibition of smooth muscle cell proliferation [17], downregulation of circulating levels of fibrinogen [18] and down regulation of endothelial cell adhesion molecules [19]. NO can also act as a chain-breaking anti-oxidant and may thereby prevent excessive oxidation of lipids in the wall of the blood vessel [20]. Most of these vasoprotective actions are attributed to tonic, physiological activity of eNOS expressed constitutively by the endothelium. There is now ample evidence to indicate that the activity of eNOS is compromized in atherosclerosis and this is supported by several observations. Endothelium-dependent, NO-mediated relaxation is significantly diminished in atherosclerotic blood vessels [21, 22] and this is compounded by the finding that organonitrate drugs such as glyceryl trinitrate and isosorbide dinitrite, long used to alleviate the symptoms of ischaemic heart disease work by releasing NO as the active metabolite [23]. It appears that several distinct mechanisms contribute to the marked impairment of endothelium-dependent, NO-mediated relaxation responses. These include, altered function of endothelial cell receptors [24], L-arginine depletion [25] and inactivation of NO by toxins associated with cigarette smoking [26]. Our own studies also illustrate that there is a marked, site specific reduction in the amounts of eNOS expressed by the endothelium overlying atherosclerotic lesions [27, 28]. By inference, reduced expression of eNOS will likely result in a marked reduction in the level of NO generated by the endothelium, consequently contributing to the diminished vascular activity of NO. Altered expression of eNOS in atherosclerosis is also supported by the finding that oxidized LDL downregulates the expression of eNOS by shortening the half-life of eNOS mRNA [29].

Although there is substantial evidence to support the involvement of eNOS in the pathogenesis of atherosclerosis, less is known about the actions of iNOS. That iNOS might be involved in atherosclerosis is suggested by the chronic inflammatory nature of the disease and the associated secretion of a wide spectrum of cytokines [30], many of which are capable of stimulating the expression of inflammatory response proteins such as iNOS. Using immunocytochemistry we have been able to show that iNOS is abundantly expressed within atherosclerotic lesions, where it was localized to macrophages, foam cells and to a lesser extent the vascular smooth muscle [31] (see Fig. 1). This finding was substantiated further by the detection of a specific 130 kDa inducible protein band in the Western blots of crude homogenates of atherosclerotic arteries. This protein correlates to the known molecular weight of iNOS and was absent from normal arteries. Moreover, iNOS mRNA was detected by *in situ* hybridization in a similar population of cells, most notably macrophages, to those shown to contain iNOS protein. Based on the known vasoprotective actions of NO it would be reasonable to postulate that expression of iNOS and activation of high-output NO synthesis would be beneficial in preventing or slowing down the progression of atherosclerosis. Indeed, as will be discussed later, iNOS has been used experimentally as a therapy to arrest the progression of vascular disease. Available evidence however, suggests that the potentially protective effects of iNOS-derived NO may be offset by the propensity for NO to react with superoxide anion within the atherosclerotic lesion resulting in the formation of peroxynitrite. Formation of peroxynitrite within cells and tissues is strongly associated with oxidative damage [12, 13]. Moreover, there is ample evidence supporting the involvement of peroxynitrite in the oxidative modification of LDL [32, 34], which as discussed above is a crucial event in the atherogenic process. Peroxynitrite may also compromise the vascular signalling activity of NO, as this NO-derived species is known to be a less potent vasodilator and platelet anti-aggregatory agent than NO itself [35].

The damaging effects associated with peroxynitrite formation are most likely mediated by its breakdown products which are formed either spontaneously or by transition metal ion catalyzed cleavage and include, nitrogen dioxide, nitronium ion and hydroxyl radical [12, 13]. In addition to oxidative activity, nitrogen dioxide and nitronium ion are strong nitrating species (targeted addition of an NO_2 group) and demonstrate an affinity for protein tyrosine residues. Nitrated tyrosine residues are considered to be a footprint or hallmark of peroxynitrite activity and they can readily be detected by chromatographic and immunocytochemical methods. The presence of nitrotyrosine has been reported in the synovial fluid of patients with

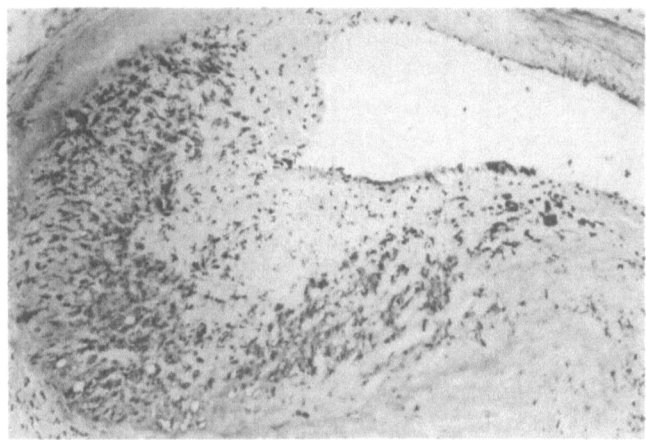

Figure 1
Frozen section from human atherosclerotic coronary artery stained with polyclonal anti-
serum to human iNOS. Numerous iNOS-positive macrophages can be seen within the lesion
and associated with the endothelium.

rheumatoid arthritis [36] and in the lung tissue of patients suffering from adult respiratory distress syndrome (ARDS) [37]. Extensive tyrosine nitration has also been described in human atherosclerosis and was present primarily in macrophages and foam cells [38]. Our studies also showed that nitrotyrosine was formed in atherosclerotic lesions and this was extended further by the illustration of the close association between iNOS and nitrotyrosine staining, both being present principally in macrophages and foam cells [31]. This led us to suggest that iNOS-derived NO is preferentially directed towards the formation of peroxynitrite within the environment of the atherosclerotic lesion and would therefore be in keeping with the earlier suggestion that the potentially beneficial effects of increased localized NO production are abrogated. In addition to immunostaining, Western blot analysis of crude homogenates of atherosclerotic arteries revealed the presence of several nitrated protein bands of varying molecular weight, which were absent from normal arteries. These nitrated protein bands also correlated closely with the protein bands detected by phosphotyrosine antibodies (see Fig. 2). The significance of this phenomenon is unclear, but it raises the possibility that peroxynitrite-mediated tyrosine nitration could interfere with tyrosine phosphorylation and thereby influence tyrosine kinase pathways, potentially resulting in the disruption of normal cell function [39].

It is of course possible that these observations on iNOS and nitrotyrosine merely reflect end stage disease-associated phenomenon. However, expression and for-

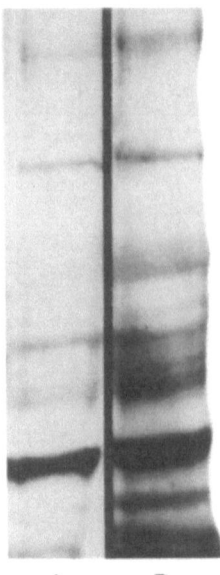

Figure 2
Western blot of crude homogenate from human atherosclerotic aorta stained (A) for nitrotyrosine and (B) phosphotyrosine. Numerous protein bands are identified with nitrotyrosine and many of these nitrated proteins correlate with proteins identified by phosphotyrosine.

A   B

mation of nitrotyrosine was observed in the earliest lesion involved in the pathogenesis of atherosclerosis – the fatty streak. This would suggest that whilst iNOS and peroxynitrite may ultimately be involved in the demise of the diseased blood vessel, these factors are also associated with the earliest origins of the disease.

Therapeutic aspects of NO and NO synthase

As discussed above, NO-donating drugs have been used successfully for many years to alleviate the symptoms of ischaemic heart disease by improving blood flow [23]. Infusion of L-arginine has also proved to be a successful strategy and presumably works by boosting endogenous NO synthesis [24, 25]. More recently, transfection of NO synthase into the wall of blood vessels after percutaneous transluminal angioplasty (PTA) has been heralded as a useful therapeutic technique to reduce the incidence of restenosis, which often ensues after PTA. PTA is now a routine procedure to revascularize stenotic blood vessels and involves directing a catheter, angiograpically, to the site of stenosis where a balloon contained within the catheter is inflated [40]. The pressure exerted by the balloon is sufficient to compress the obstructing lesion, dilating the lumen of vessel and thereby improving blood flow. The major failing of this technique however, is the high incidence of restenosis and this process is related primarily to the fact that the procedure of inflating the balloon often damages or denudes the endothelium [41, 42]. Loss of the antithrombotic surface provided by the endothelium can provide a nidus for clot formation. Further, mitogenic factors released by damaged endothelium and aggregating platelets, in particular platelet-derived growth factor (PDGF) can contribute to the accelerated intimal proliferative response [43]. Animal models have illustrated that iNOS is expressed by vascular smooth muscle after PTA, and this has been interpreted as a protective response to counteract the loss of endothelial-derived NO [44]. Expression of iNOS after PTA in this model however, was only transient and disappeared after a few days. This might relate to the actions of mitogens like PDGF or blood coagulation products, in particular thrombin, both of which are known to down-regulate iNOS [45, 46]. In this situation, boosting the activity of vasculoprotective genes would likely prove to be beneficial and this is currently being addressed by the emerging technology of vascular gene transfection [47]. This technique has in fact now been used to introduce eNOS [48] and iNOS [49, 50] into human blood vessels and both have proved successful in limiting the extent of intimal proliferation after PTA.

The protective effects afforded by introducing iNOS into the wall of a blood vessel appear to contradict its damaging actions described in atherosclerosis. This might be explained by the fact that the localized expression and activity of iNOS is not damaging *per se*, rather it is the nature of the environment into which NO is released that modifies its actions. In the case of atherosclerosis, the presence of superoxide anions promotes the deleterious formation of peroxynitrite and thus, methods to prevent formation of this damaging oxidant would likely be beneficial. To this end, targeted addition of superoxide dismutase into atherosclerotic rabbit blood vessels has proved to be successful in restoring endothelium-dependent relaxation [51]. Furthermore, there are numerous other studies describing the preservation of endothelium-dependent vasodilatation by various antioxidant agents such as vitamin E and oestrogen [52].

Cyclooxygenase(s)

Prostaglandins are a family of lipid mediators that are enzymatically made from arachidonic acid. Cyclooxygenase (COX) converts arachidonic acid to prostaglandin endoperoxide (PGH_2) and PGH_2 serves as a substrate for various synthases to produce active metabolites, which include prostacyclin (PGI_2), prostaglandins E1 and E2 and thromboxane [53]. There are two distinct isoforms of COX – COX-1 and COX-2. COX-1 is a constitutively expressed enzyme and is present in virtually all cells and tissues, including vasculature. In a manner similar to the constitutive NO synthases, the activity of COX-1 is generally associated with low level production of prostaglandins, which exert a tonic, physiologic effect. By contrast, COX-2 is generally absent from normal healthy cells and tissues, however, in response to various inflammatory stimuli the expression of this enzyme can be induced in numerous cells types but primarily in macrophages, fibroblasts, endothelial cells and smooth muscle cells. The activity of this enzyme accounts for the release of large quantities of prostaglandins at sites of inflammation. In this respect, COX-2 behaves in a manner similar to iNOS. Indeed, both COX-2 and iNOS are often expressed in the same inflammatory diseases and as will be discussed later there is now ample evidence supporting interactions between the iNOS and COX-2 pathways.

COX and atherosclerosis

Perhaps the most important COX metabolite in terms of vascular function is PGI_2, and PGI_2 is considered to be the main product of arachidonic acid metabolism in the majority of vascular tissues [54]. The principal activity of PGI_2 is inhibition of platelet aggregation and to a lesser extent inhibition of platelet-endothelial cell adhesion [55]. PGI_2 is also a vasodilator agent. In this respect, PGI_2 exerts a vasoprotective, antithrombotic and antiatherogenic effect in a manner similar to NO. This is further emphasized by the finding that PGI_2 and NO work synergistically to prevent platelets aggregating and to disperse existing platelet aggregates [14].

Reduced activity of PGI_2 in atherosclerotic arteries has been reported by some investigators [56, 57], whilst others have indicated that PGI_2 synthesis is increased [58, 59]. In support of the increased synthesis of PGI_2 preliminary studies made in our laboratory have indicated that expression of COX-2 is induced in human atherosclerotic arteries [60]. COX-2 was localized to endothelial cells, macrophages, foam cells and vascular smooth muscle. This localization is almost entirely co-incident with that of iNOS (see Fig. 3) and is in keeping with the notion that both COX-2 and iNOS are proteins that are co-expressed as part of the integrated response to an inflammatory stimulus.

The function of COX-2 in atherosclerosis is at present unclear but it may fulfil a protective role. This is suggested by several observations. Lysophosphatidylcholine, a component of oxidized LDL, has been shown to induce expression of COX-2 and this was interpreted as a protective response to the damaging effects of oxidized LDL [61]. Another study has illustrated that COX-2 (and also eNOS) are upregulated by laminar flow shear stress, whereas turbulent flow or low shear stress does not upregulate these genes [62]. The significance of this observation on hemodynamic forces and their involvement in the development of atherosclerosis relate to the fact that regions of laminar shear stress are usually "lesion protected" areas whereas regions of disturbed, turbulent flow, often associated with blood vessel bifurcations are "lesion prone" areas. This study also found that superoxide dismutase was upregulated in these "lesion protected" areas which further supports the work discussed above on the use of antioxidants to preserve and promote the actions of vascular mediators with antiatherogenic potential. PGI_2, PGE_1, and PGE_2

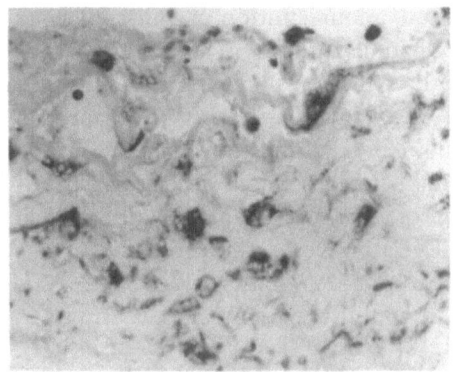

Figure 3
Frozen section from human atherosclerotic coronary artery stained (a) with a polyclonal antiserum to human iNOS and (b) with a polyclonal antiserum to COX-2. Staining for iNOS and COX-2 is seen in lipid-laden foam cells.

are also implicated in stimulating fibrinolysis [63–65]. In this respect prostaglandins are once again protective, as impaired fibrinolysis is strongly associated with the development of atherosclerosis. In addition, prostaglandins of the E series, which are often released by macrophages exert vasodilator effects and synergise with NO to prevent platelet aggregation [66].

Therapeutic aspects of COX and prostaglandins

There are several therapeutic benefits from administration of prostacyclin and its stable analogues such as cicaprost and iloprost. Beneficial effects in the treatment of atherosclerosis include decreased activation of platelets and improvement of impaired vasodilatation [67]. Another PGI_2 analogue TFC-132, has also proved to be successful in limiting the extent of neointimal thickening after mechanically

117

induced injury to rabbit aorta [68]. Furthermore, administration of PGI_2, PGE_1 or isradipine, a stimulator of endogenous prostaglandin synthesis, in animal models or human subjects improved lipid metabolism in the arterial wall resulting in a marked decrease in potentially damaging cholesterol esters [69]. Aspirin is also a proven treatment for ischaemic heart disease [70]. On the basis of the above positive effects of prostaglandins in alleviating the symptoms associated with atherosclerosis, this is a curious observation since aspirin is a non-steroidal anti-inflammatory drug which non-selectively inhibits both COX-1 and COX-2. However, the protective effects associated with taking aspirin have been ascribed to the inhibition of thromboxane, which is a potent activator of platelet aggregation and vasoconstriction [70].

Interactions between COX and NO synthase

Increasing evidence suggests that there is considerable "cross-talk" between COX and NO synthase. NO has been shown to enhance the activity and increase the formation of prostaglandins in murine recombinant COX-1 and COX-2 and in RAW macrophages [71]. More recently NO was shown to enhance the release of PGI_2 from endothelial cells and by this mechanism NO and NO stimulated PGI_2 synthesis, contribute to the marked anti-platelet effects associated with the administration of nitrovasodilators [72]. This observation is also in keeping with earlier work illustrating synergy between the vascular actions of NO and PGI_2. The mechanism of NO stimulated COX synthesis is not clearly defined although one study has indicated that NO activates COX directly by binding to its heme prosthetic group [71]. Others however, have suggested that NO activates COX by nitrosylation of critical amino acid residues in the active site [73].

There is also substantial evidence showing that iNOS and COX-2 are induced concurrently in a number of models of inflammation including rabbit hydronephrotic kidney endotoxin-induced septic shock and carrageenan-induced pouch and paw inflammation [74, 75]. Our own studies have also suggested that both iNOS and COX-2 are induced in human atherosclerosis. In another separate study we showed that iNOS and COX-2 are induced in osteoblast cell cultures by pro-inflammatory cytokines [76]. Moreover, using this model we also found that NO not only stimulated prostaglandin synthesis, but also upregulated the expression of COX- 2 protein. COX activity also provides a source of superoxide anion, raising the potential for localized formation of peroxynitrite and there is now evidence to show that peroxynitrite can influence the activity of prostaglandins. For example, a recent study demonstrated that peroxynitrite could block the activity of PGI_2 synthase and thereby attenuate the production of PGI_2 [77]. Peroxynitrite can also oxidatively modify the COX substrate arachidonic acid, yielding F2-isoprostanes [78, 79]. F2-isoprostanes have been described as novel bioactive prostaglandin F2-like compounds, which exert powerful vasconstrictor effects and thus might contribute to the impaired vascular function associated with atherosclerosis. There is also one study demonstrating downregulation of iNOS expression by prostaglandin E2 [80], suggesting that the basis for the interaction between metabolites of the NO synthase and COX pathways is not solely dependent on NO driven regulation of COX.

Summary

In this short review we have discussed the contribution made by NO and prostaglandins to the regulation of atherogenesis, focusing in particular on the function of iNOS and COX-2. There is certainly evidence to indicate that iNOS and COX-2 are expressed in atherosclerosis and this might represent a protective response. Indeed, therapies involving restoration or boosting the vascular activity of NO synthase and COX metabolites have a proven beneficial effect in alleviating or retarding the clinical symptoms and pathology associated with atherosclerosis. It seems, however, that other factors associated with atherosclerosis, notably reactive oxygen intermediates, have the capacity to modify in a deleterious manner, through the formation of peroxynitrite, the activity of these enzymatic pathways, in particular iNOS. Formation of peroxynitrite not only comprises the vascular signalling activity of NO it also promotes oxidative damage. Moreover, through the known effects of NO on the stimulation of COX activity it is possible that altered NO-dependent signalling may also have a knock-on effect on the synthesis and activity of COX metabolites. Taken together, these observations add to our understanding of the aetiology and pathogenesis of atherosclerosis and may also help in devising new strategies to treat or prevent this disease.

References

1   Ross R (1993) Atherosclerosis: A defence mechanism gone awry. *Am J Path* 143: 987–1002
2   O'Brien KD, Chait A (1994) The biology of the artery wall in atherogenesis. *Med Clin North Am* 78: 41–67
3   Moncada S, Gryglewski R, Bunting S, Vane JR (1976) An enzyme isolated from arteries transforms prostaglandin endoperoxides to an unstable substance that inhibits platelet aggregation. *Nature* 263: 663–665
4   Palmer RMJ, Ferrige AG, Moncada S (1987) Nitric oxide release accounts for the biological activity of endothelium-derived relaxing factor. *Nature* 327: 524–526
5   Brown MS, Goldstein JL (1983) Lipoprotein metabolism in the macrophage: Implications for cholesterol deposition in atherosclerosis. *Ann Rev Biochem* 52: 223–250
6   Mitchinson MJ, Ball RY (1987) Macrophages and atherogenesis. *Lancet* 2: 146–148

7   Ross R, Glomset JA (1976) The pathogenesis of atherosclerosis. *N Eng J Med* 295: 369–377

8   Ross R (1986) The pathogenesis of atherosclerosis – an update. *N Eng J Med* 314: 488–500

9   Knowles RG, Moncada S (1995) Nitric oxide synthases in mammals. *Biochem J* 298: 249–258

10 Gross SS, Wolin MS (1995) Nitric oxide: Pathophysiological mechanisms. *Ann Rev Physiol* 57: 737–769

11 Nussler AK, Billiar TR (1993) Inflammation, immunoregulation and inducible nitric oxide synthase. *J Leuko Biol* 54: 171–178

12 Crow JP, Beckman JS (1995) Reactions between nitric oxide, superoxide and peroxynitrite: Footprints of peroxynitrite *in vivo*. *Adv Pharmacol* 34: 17–44

13 Freeman BA (1994) Free radical chemistry of nitric oxide. *Chest* 105 (Suppl): 79S–80S

14 Radomski MW, Palmer RMJ, Moncada S (1987) The anti-aggregating properties of vascular endothelium: interactions between prostacyclin and nitric oxide. *Br J Pharmacol* 92: 639–646

15 Radomski MW, Palmer RMJ, Moncada S (1987) Endogenous nitric oxide inhibits human platelet adhesion to vascular endothelium. *Lancet* 2: 1057–1058

16 Kubes P, Suzuki M, Granger DN (1991) Nitric oxide: An endogenous mediator leucocyte adhesion. *PNAS* 88: 4651–4655

17 Garg UC, Hassad A (1989) Nitric oxide-generating vasodilators and 8-bromo-cGMP inhibit mitogenesis and proliferation of cultured rat vascular smooth muscle cells. *J Clin Invest* 83: 1774–1777

18 Kawabata A (1996) Evidence that endogenous nitric oxide modulates plasma fibrinogen levels in the rat. *Br J Pharmacol* 117: 236–237

19 Khan BV, Harrison DG, Olbrych MT, Alexander RW, Medford RM (1996) Nitric oxide regulates vascular cell adhesion molecule 1 gene expression and redox-sensitive transcriptional events in human vascular endothelial cells. *PNAS* 93: 9114–9119

20 Hogg N, Kalyanaraman B, Joseph J, Struck A, Parthasarathy S (1993) Inhibition of low-density lipoprotein oxidation by nitric oxide. Potential role in atherogenesis. *FEBS Lett* 334: 170–174

21 Forstermann U, Mugge A, Alheid U, Haverich A, Frolich JC (1988) Selective attenuation of endothelium-mediated vasodilation in human atherosclerotic arteries. *Circ Res* 62: 185–190

22 Chester AH, O'Neil GS, Moncada S, Tadjkarimi S, Yacoub MH (1991) Low basal and stimulated relaese of nitric oxide in atherosclerotic epicardial coronary arteries. *Lancet* 336: 897–900

23 Loskove JA, Frishman WH (1995) Nitric oxide donors in the treatment of cardiovascular and pulmonary diseases. *Am Heart J* 129: 604–613

24 Drexler H, Zeiher AM, Meinzer K, Just H (1991) Correction of endothelial dysfunction in coronary microcirculation of hypercholesterolaemic patients by L-arginine. *Lancet* 338: 1546–1550

protein modifications on tyrosine phosphorylation and degradation. *FEBS Lett* 385: 63–66

40 Gruentzig AR, Myler RK, Hanna EH, Turina MI (1977) Restenosis after balloon angioplasty. *Circulation* 84 (Suppl2): II55–56

41 Levy, RI, Mock MB, Willman VL, Passamani ER, Fromer PL (1981) Percutaneous coronary angioplasty – a status report. *N Eng J Med* 305: 399–400

42 Nobuyoshi M, Kimura T, Ohishi H, Horiuchi H, Nosaka H, Hamasaki N, Yoki H, Kim K (1991) Restenosis after percutaneous transuluminal angioplasty: Pathologic observations in 20 patients. *J Am Coll Cardiol* 17: 433–440

43 Wilcox JN (1993) Molecular biology: Insight into the causes and prevention of restenosis after arterial intervention. *J Am Coll Cardiol* 72: 88E–95E

44 Douglas SA, Vickery-Clarke LM, Ohlstein EH (1994) Functional evidence that balloon angioplasty results in transient nitric oxide synthase induction. *Eur J Pharmacol* 255: 81–89

45 Schini VB, Durante DL, Elizondo E, Scott-Burden T, Janquero DL, Schafer AI, Vanhoutte PM (1992) The induction of nitric oxide synthase activity is inhibited by TGF-b1, PDG-FAB, and PDGFBB in vascular smooth muscle cells. *Eur J Pharmacol* 216: 379– 383

46 Schini VB, Catovsky S, Durante DL, Scott-Burden T, Schafer AI, Vanhoutte PM (1993) Thrombin inhibits induction of nitric oxide synthase in vascular smooth muscle cells. *Am J Physiol* 264: H611–H616

47 Nabel EG, Plaitz, Boyce FM, Tranley JC. Nabel GC (1989) Recombinant gene expression *in vivo* within endothelial cells of the arterial wall. *Science* 244: 1342–1344

48 Von der Leyen HE, Gibbons GH, Moishita R, Lewis NP, Zhang L, Nakajima M, Kaneda Y, Cooke JP, Dzau VJ (1995) Gene therapy inhibiting neointimal vascular lesion: *In vivo* transfer of endothelial cell nitric oxide synthase gene. *PNAS* 92: 1137–1141

49 Tzeng E, Shears LL, Lotze MT, Billiar TR (1996) Gene therapy. *Curr Probl Surg* 33: 961–1041

50 Shears LL, Kawaharada N, Tzeng E, Billiar TR, Watkins SC, Kovesdi I, Lizonova A, Pham SM (1997) Inducible nitric oxide synthase supppresses the developmeent of allograft arteriosclerosis. *J Clin Invest* 100: 2035–2042

51 Mugge A, Ewell JH, Peterson TE, Hofmeyer TG, Heistad DD, Harrison DG (1991) Chronic treatment with polyethylene-glycolated superoxide dismutase partially restores endothelium-dependent vascular relaxations in cholesterol-fed rabbits. *Circ Res* 69: 1293–1300

52 Keaney JJr, Vita JA (1995) Atherosclerosis, oxidative stress, and antioxidant protection in endothelium-derived relaxing factor action. Progress in Cardiovasc Dis 33: 129–154

53 Wu KK (1995) Inducible cyclooxygenase and nitric oxide synthase. *Adv Pharmacol* 33: 179–207

54 Bunting S, Gryglewski R, Moncada S, Vane JR (1976) Arterial walls generate from prostaglandin endoperoxides a substance (prostaglandin x) which relaxes strips of mesenteric and coeliac arreries and inhibits platelet aggregation. *Prostaglandins* 12: 897–913

25 Cooke JP, Andon NA, Girerd X, Hirsch A, Craeger MA (1991) L-arginine restores cholinergic relaxation of hypercholesterolaemic rabbit throacic aorta. *Circulation* 83: 1057–1062

26 Celermajer DS, Adams MR, Clarkson P, Robinson J, McCredie R, Donald A, Deanfield JE (1996) Passive smoking and impaired endothelium-dependent arterial dilatation in healthy young adults. *N Eng J Med* 334: 150–154

27 Buttery LDK, Chester AH, Springall DR, Borland JAA, Michel T, Yacoub MH, Polak JM (1996) Explanted vein grafts with an intact endothelium demonstrate reduced focal expression of endothelial NO synthase specific to atherosclerotic sites. *J Path* 179: 197–203

28 Buttery LDK, Polak JM (1995) Localization of nitric oxide synthase: alterations in disease. *Curr Diag Path* 2: 111–121

29 Liao JK, Shin WS, Lee WY, Clark SL (1995) Oxidized low-density lipoprotein decreases the expression of endothelial NO synthase. *J Biol Chem* 270: 391–324

30 Libby P, Hansson GK (1991) Involvement of the immune system in human atherogenesis: Current knowledge and unanswered questions. *Lab Invest* 64: 5–14

31 Buttery LDK, Springall DR, Chester AH, Evans TJ, Parums, Standfield N, Yacoub MH, Polak JM (1996) Inducible NO synthase is present within atherosclerotic lesions and promotes the formation and activity of peroxynitrite (1996) *Lab Invest* 75: 76–78

32 Darley-Usmar VM, Hogg N, O'Leary VJ, Tsai M (1992) The simultaneous generation of superoxide and nitric oxide can initaite lipid peroxidation in human low-density lipoprotein. *Free Rad Res Comm* 17: 19–20

33 Graham A, Hogg N, Kalyanaraman B, O'Leary VJ, Darley-Usmar V, Moncada S (1993) Peroxynitrite modification of low density lipoprotein leads to recognition by the macrophage scavenger receptor. *FEBS Lett* 330: 181–185

34 White CR, Broack TA, Chang LY, Crapo J, Briscoe P, Ku D, Bradley WA, Gianturco SH, Gore J, Freeman BA, Tarpey MM (1994) Superoxide and peroxynitrite in atherosclerosis. *PNAS* 91: 1044–1048

35 Moro MA, Darley-Usmar VM, Goodwin DA, Read NG, Zamora-Pino R, Feelisch M, Radomski M, Moncada S (1994) Paradoxical fate and biological action of peroxynitrite on human platelets. *PNAS* 91: 6702–6706

36 Kaur H, Halliwell B (1994) Evidence for nitric-oxide mediated oxidative damage in chronic inflammation – Nitrotyrosine in serum and synovial fliud from rheumatoid patients. *FEBS Lett* 350: 9–12

37 Haddad IY, Pataki G, Hu P, Galliani C, Beckman JS, Mataon S (1994) Quantification of nitrotyrsosine levels in lung sections of patients and animals with acute lung injury. *J Clin Invest* 94: 2407–2413

38 Beckman JS, Ye YZ, Anderson PG, Chen J, Accavitti MA, Tarpey MM, White CR (1994) Extensive nitration of protein tyrosine residues in human atherosclerosis detected by immunohistochemistry. *Biol Chem-Hoppe Seyer* 375: 81–88

39 Gow A, Duran D, Malcolm S, Ischiropoulos H (1996) Effects of peroxynitrite-induced

55 Moncada S, Vane JR (1979) Pharmacology and endogenous roles of prostaglandin endoperoxides, thromboxane A2 and prostacyclin. *Pharmacol Rev* 30: 293–331

56 Dembinska-Kiec A, Gryglewska T, Zmuda A, Gryglewski RJ (1977) The generation of prostacyclin by arteries and by the coronary vascular bed is reduced in experimental atherosclerosis in rabbit. *Prostaglandins* 14: 1025–1034

57 D'Angelo V, Villa S, Musliwiec M, Donati MB, De Gaetano G (1978) Defective fibrinolytic and prostacyclin like activity in human atheromatous plaques. *Thromb Haemostasis* 39: 535–536

58 Tremoli E, Socini A, Petroni A, Galli C (1982) Increased platelet aggregability is associated with increased prostacyclin production by vessel walls in hypercholesterolaemic rabbits. *Prostaglandins* 24: 397–404

59 Fitzgerald GA, Smith B, Pedersen AK, Brash AR (1984) Increased prostacyclin biosynthesis in patients with severe atherosclerosis and platelet activation. *N Eng J Med* 310: 1065–1068

60 Baker CSR, Hall RJC, Evans TJ, Pomerance A, Maclouf J, Creminon C, Yacoub MM, Polak JM (1998) Cyclooxygenase-2 is widely expressed in atherosclerotic lesions affecting native and transplanted human coronary arteries and co-localizes with inducible nitric oxide synthase and nitrotyrosine particularly in macrophages. *Arterioscler Thromb Vasc Biol; in press*

61 Zembowicz A, Jones SL, Wu KK (1995) Induction of cyclooxygenase 2 in human umbilical vein endothelial cells by lysophosphatidylcholine. *J Clin Invest* 96: 1688

62 Topper JN, Cai J, Falb D, Gimbrone MA Jr (1996) Identification of vascular endothelial genes differentially responsive to fluid mechanical stimuli: Cyclooxygenase-2 , manganese superoxide dismutase, and endothelial cell nitric oxide synthase are selectively up-regulated by steady laminar shear stress. *PNAS* 93: 10417–10422

63 Crutchley DJ, Conanan LB, Maynard JR (1982) Stimulation of fibrinolytic activity in human skin fibroblasts by prostaglandins E1, E2 and I2. *J Pharm Exp Therap* 222: 544–549

64 Dembinska-Kiec A, Kostka-Trabska E, Gryglewski RJ (1982) Effect of prostacyclin on fibronolytic activity in patients with arteriosclerosis obliteran. *Thromb Haemost* 47: 90

65 Szczeklik A, Kopec M, Sladek K (1983) Prostacyclin and the fibrinolytic system in ischaemic vascular disease. *Thromb Res* 29: 655–660

66 Kadish J (1995) Endothelium, fibrinolysis, cardiac risk factors and prostaglandins: a unified model of atherogenesis. *Med Hypoth* 45: 205–213

67 Braun M, Sarbia M, Kienbaum P, Hohlfels T, Weber A, Schror K (1992) Anti-atherosclerotic properties of oral cicaprost in hypercholesterolaemic rabbits. *Agents and Actions* 37: 282–288

68 Asada Y, Kisanuki A, Hatakeyama K, Takahama S, Koyama T, Kurozumi S, Sumiyoshi A (1994) Inhibitory effects of prostacyclin analogue, TFC-132, on aortic neointimal thickening *in vivo* and smooth muscle cell proliferation *in vitro*. *Prostaglandin Leukot Essent Fatty Acids* 51: 245–248

69 Sinzinger H, Rogatti W (1994) Prostaglandins and arterial wall lipid metabolism *in vitro*, *ex vivo* and *in vivo* radiographic studies. *J Physiol Pharmacol* 45: 27–40

70 Steering Committe of the Physician's Health Study Research Group (1989) Final report on the aspirin component of the ongoing Physician's Health Study. *N Eng J Med* 321: 129

71 Salvemini D, Misko TP, Masferrer JL, Siebert K, Currie MG, Needleman (1993) Nitric oxide activates cyclooxygenase enzymes. *PNAS* 90: 7240–7244

72 Salvemini D, Currie MG, Mollace V (1996) Nitric oxide mediated cyclooxygenase activation. A key event in the antiplatelet effects of nitrosovasodilators. *J Clin Invest* 97: 2562–2568

73 Hajjar DP, Lander HM, Pearce FS, Upmacis RK, Pomerantz KB (1995) Nitric oxide enhances prostaglandin H synthase activity by a heme-independent mechanism: evidence implicating nitrosothiols. *J Am Chem Soc* 117: 3340–3346

74 Mitchell JA, Larkin S, Williams TJ (1995) Cyclooxygenase 2: Regulation and relevance in inflammation. *Biochem Pharamcol* 50: 1535–1542

75 Salvemini D, Masferrer JL (1996) Interactions of nitric oxide with cyclooxygenase: *In vitro*, *ex vivo* and *in vivo* studies. *Met Enzymol* 269: 12–25

76 Hughes FJ, Buttery LDK, Hukkanen MVJ, O'Donnell A, Polak JM (1998) Cytokine-induced prostaglandin E_2 synthesis and cyclo-oxygenase-2 activity are regulated both by a nitric oxide-dependent and -independent mechanism in rat osteoblasts *in vitro*. *J Biol Chem*; *in press*

77 Zou M, Martin C, Ullrich V (1997) Tyrosine nitration as a mechanism of selective inactivation of prostacyclin synthase by peroxynitrite. *Biol Chem* 378: 707–713

78 Lynch SM, Morrow JD, Roberts LJ II, Frei B (1994) Formation of non-cyclooxygenase-derived prosatnoids (F2-isoprostanes) plasma and low density lipoprotein exposed to oxidative stress *in vitro*. *J Clin Invest* 93: 998–1004

79 Moore KP, Darley-Usmar V, Morrow J, Roberts LJ (1995) Formation of F2-isoprostanes during oxidation of human low-density lipoprotein and plasma peroxynitrite. *Circ Res* 77: 335–341

80 Milano S, Arcoleo F, Dieli M, D'Agostino R, D'Agostino P, De Nucci G, Cillari E (1995) Prostaglandin E2 regulates inducible nitric oxide synthase in the murine macrophage cell line J774. *Prostaglandins* 49: 105–115

The role of the inducible enzymes cyclooxygenase-2, nitric oxide synthase and heme oxygenase in angiogenesis of inflammation

Michael P. Seed[1], Derek Gilroy, Mark Paul-Clark, Paul R. Colville-Nash, Dean Willis, Annette Tomlinson and Derek A. Willoughby

Experimental Pathology, William Harvey Research Institute, Saint Bartholomew's and the Royal London School of Medicine and Dentistry, Charterhouse Square, London EC1M 6BQ, UK
[1]present address: Paneutics Ltd., P.O. Box 1358, Swindon SN3 4GP, UK

Introduction to angiogenesis

Angiogenesis, the formation of new blood vessels, is an essential part of the body's physiology. In the non-pathological state, the process is largely quiescent and endothelial cell turnover may be measured in terms of years. However, it is an essential component of a variety of normal functions such as embryogenesis, normal tissue growth and the menstrual cycle. It also plays a role in the pathology of a variety of disease states and this led Judah Folkman, who many consider to be the "father of angiogenesis research", to coin the term "angiogenesis-dependent disease" [1]. Some of the obvious examples which may be considered to fall into such a disease categorisation include neovascular glaucoma, hemangiomas and other tumors which need vascular support for their tissue expansion and metastatic activity. However, a number of other pathologies exist in which angiogenesis is a prominent feature; these include many of the chronic inflammatory diseases such as rheumatoid arthritis [2]. In these latter diseases, the neovasculature not only acts as a route for the increased nutrient supply required by the developing tissue, but also provides a greatly exaggerated area of activated endothelium which transmits pro-inflammatory signals, as well as receiving them, and allows the recruitment of large numbers of inflammatory leucocytes. The processes and cytokines involved in this proliferative capillary response have been recently reviewed [3, 4, 5]. The modulation of angiogenesis in inflammation therefore holds great therapeutic promise for the treatment of chronic inflammatory disease.

Angiogenesis has for a long time been assumed to be a requirement for the continued development of chronic inflammation, but this has been difficult to prove through pharmacological means. Studies using angiostatic steroids (without intrinsic antiinflammatory activity at the doses used) have shown that the development of irritant and delayed type hypersensitivity granulomas can be inhibited by the restriction of angiogenesis [6]. On the other hand, stimulation of angiogenesis results in the acceleration of granulomatous tissue development, leucocyte recruitment and cartilage degradation in the cotton/cartilage model of arthritis [4], whilst its inhibi-

Inducible Enzymes in the Inflammatory Response, edited by D.A. Willoughby and A. Tomlinson

tion has the desired opposite effects [4, 7], reducing granulomatous tissue development and maintaining cartilage viability for longer. Also, using another angiostatic steroid, methoxyoestradiol, the development of collagen II arthritis [8] can be ameliorated, and the modern angiostatic agent TNP-470 (a synthetic derivative of the angiostatic antibiotic fumagillin) is effective in inhibiting both adjuvant [9] and collagen II arthritis [10]. Various anti-rheumatic drugs have also been shown to be capable of reducing angiogenesis [11, 12] and interestingly, early work in the neoplastic field indicated that certain non-steroidal anti-inflammatory drugs may have the same effect [13, 14], although the assays were primitive by present standards.

These latter studies have prompted the investigation of the role of mediator enzymes such as cyclooxygenase in inflammatory angiogenesis.

The process of angiogenesis in inflammation is complex. The proliferative response of capillaries to inflammatory stimuli begins at the very earliest of stages with the exudation of plasma during capillary permeabilisation. Activation of the endothelium through cytokines such as interleukin-1 (IL-1) and tumour necrosis factor α (TNFα) produces an environment which is pro- rather than anti-coagulant [15, 16]. Plasmin production is shifted towards the basement membrane, and the endothelial cells retract. This allows oedema formation, fibrin deposition and endothelial cell as well as pericyte migration, the first step towards angiogenesis. The endothelial cells at the tip of the newly forming capillary migrate through the extracellular matrix, with the following cells dividing and then laying down basement membrane and lumenating. A capillary is fully formed when two sprouts join to form a capillary loop. This is then repeated to produce capillary arcades (see
). It is now recognised that this process is finely choreographed through the influence of inflammatory mediators, cytokines, growth factors and soluble adhesion molecules. The redundancy and interplay between these factors make the mapping of this network and elucidation of their relative roles and importance a daunting task.

Many mediators have been noted to influence angiogenesis in chronic inflammation. For example, angiogenin, basic fibroblast growth factor (FGF-2) [17], TNFα, IL-8, platelet activating factor (PAF) [18], IL-1, transforming growth factor β (TGFβ) [19], vascular endothelial cell growth factor (VEGF) [20], endothelin-1 [21] and substance P [22] have all been implicated in inflammatory angiogenesis linked to rheumatoid arthritis. Whilst VEGF selectively targets the endothelium,

Figure 1

Key steps in the angiogenic response. Activation of the endothelium (1) leads to elaboration of proteases and degradation of the basement membrane (2), migration (3) and proliferation (4) of endothelial cells to form new vascular sprouts which form a new lumen and synthesise a new basement membrane (5), join together and form capillary loops (6) from which new generations of vessels can be formed (7).

1) Activation of endothelial cells

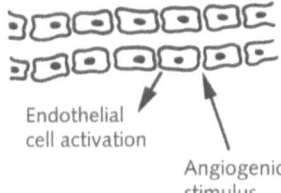

Endothelial
cell activation

Angiogenic
stimulus

5) Lumenation of sprout
and synthesis of new basement membrane

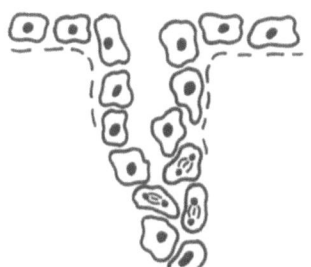

2) Secretion of proteases to degrade
basement membrane and tissue

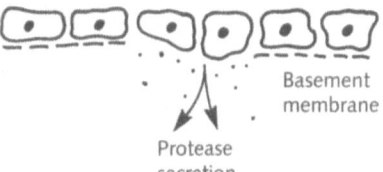

Basement
membrane

Protease
secretion

6) Two sprouts link to from capillary loop

3) Formation of capillary sprout

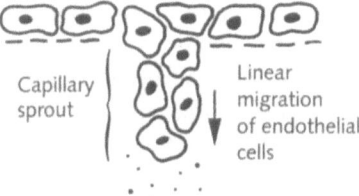

Capillary
sprout

Linear
migration
of endothelial
cells

7) Development of second
generation capillary sprouts

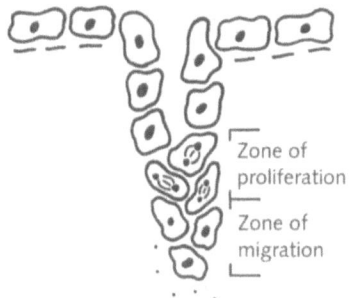

4) Growth of capillary sprout

Zone of
proliferation

Zone of
migration

most other mediators may act both directly and indirectly (via altering the activity of cells other than the endothelium) on the angiogenic process. Their contribution to angiogenesis *in vivo* is thus difficult to assess, particularly in an inflammatory enviroment. However, many act through a complex signalling hierarchy in which nitric oxide (NO) and/or prostaglandin synthesis have been shown to be important components. These are synthesised by the various enzyme isoforms of nitric oxide synthase (NOS) and cyclooxygenase (COX) respectively.

Relationship between the inducible enzymes and the vasculature in inflammation

The inducible enzymes which are the subject of this chapter synthesize substances which are mediators and modulators of the inflammatory process and their expression is closely associated with the vasculature. The products of inducible cyclooxygenase (COX-2) and inducible nitric oxide synthase (iNOS) are prostaglandins (PGs) and NO respectively. These molecules have been ascribed both pro- and anti-inflammatory roles and have profound effects on vascular function [23, 24]. Heme oxygenase (HO) catabolizes the breakdown of heme to biliverdin (which is subsequently converted enzymatically to the potent free radical scavenger bilirubin) free iron and carbon monoxide (CO) [25], which like NO acts as a vasodilator [26]. The activity of the inducible isoform HO-1 has been coincident with the suppression of inflammatory reactions [27] and much interest is now being expressed in the role of this enzyme and its products in the inflammatory response and its potential as a therapeutic target (see Chapter by D. Willis for further details).

The protein expression of COX-2, iNOS and HO-1 has been widely demonstrated in animal models of inflammation and human inflammatory disease. We have shown COX-2 expression in neutrophils and macrophages in the rat carrageenin-induced pleural model of acute inflammation [28]; in endothelial cells of venules and newly formed capillaries as well as infiltrating macrophages and fibroblasts in chronic inflammation in the murine chronic granulomatous tissue air pouch [29]; and in endothelial cells, neutrophils, macrophages and fibroblasts in the hyperplastic synovium of the rat monoarticular arthritis. In human rheumatoid pannus, the pattern is similar, with COX-2 expressed in endothelial cells, infiltrating monocytes and deeper synovial fibroblasts [30].

The pattern of iNOS expression in models of acute and chronic inflammation is similar to that of COX-2 [31]. In the rheumatoid joint, iNOS is expressed in the synovial lining layer, sub-synovium, CD68+ macrophages, fibroblasts, vascular smooth muscle, and in chondrocytes but is absent from neutrophils [32, 33]. Both the constitutively expressed isoforms of HO, HO-2, and HO-1 have been shown in many reports to be present in endothelial cells [34, 35]. Studies in our laboratories have shown HO-1 expression in macrophages in the rat carrageenin-induced plerisy, in neutrophils in the Arthus reaction, and in macrophages and vascular smooth muscle in endotoxic rats and chronic granulomatous tissue.

The remainder of this chapter will focus on the role of these inducible enzymes in angiogenesis.

Cyclooxygenase, prostaglandins and angiogenesis

The prostaglandins possess the most clear angiogenic effects. PGE_2 is pro-angiogenic in the rabbit cornea [36] and introduction of PGE_1 into this model and the chick chorioallantoic membrane (CAM) assay induces capillary growth [37, 38]. Further, the application of PGE_2 or PGE_1 to rat femoral veins *in vivo* also induces intense capillary sprouting [39]. Stable prostacyclin analogues have similarly been shown to be pro-angiogenic in the CAM assay [40] and the topical administration of isoprostenol to diabetic mice also induces angiogenesis and accelerates wound healing [41].

In other cells important in controlling tissue responses to inflammatory stimuli, mediators such as the cytokine IL-1 induce macrophages to release PGE_2 [42] and endothelial cells to release PGI_2 [43] which is enhanced in the presence of other angiogenic growth factors such as endothelial cell growth factor [44]. VEGF has a similar effect in inducing PGE_2 release although this may be secondary to initial elaboration of NO and FGF-2 [45] (see below). Interestingly both VEGF and COX-2 are expressed in rheumatoid synovia by macrophages and fibroblasts surrounding microvessels [46–48], which themselves express the inducible isoform of COX [30]. Furthermore, a positive feedback loop may exist in this sequence as PGE_2 can induce the expression of VEGF in synovial fibroblasts [49].

In vivo, in most of these situations it is difficult to separate the effects of the prostaglandins from inflammatory cell recruitment and thus indirect stimulation of angiogenesis by angiogenic factors released from these cells. There are few positive reports of prostaglandin effects directly on endothelial cell angiogenic functions *in vitro*. However, whilst there is little evidence for PGE stimulating the proliferation of endothelial cells, it does induce endothelial cell migration [36]. Similarly the endothelial cell chemotactic effect of thrombin is reported to be mediated through prostaglandins [44].

Nitric oxide synthase, nitric oxide and angiogenesis

In the case of NO, the role of this mediator in angiogenesis is less clear and appears sensitive to both the assays and the concentrations used. NO donors promote endothelial cell proliferation and migration both *in vivo* and *in vitro*, whilst NOS inhibition suppresses these events [45, 50]. Elaboration of urokinase type plasminogen activator (a key protease in the initiation of the angiogenic process) by endothelial cells is also enhanced by incubation of the cells with the NO donor sodium nitroprusside [51]. This increase was blocked by coincubation with blocking antibodies to FGF-2 suggesting an indirect mechanism of action via NO enhancing FGF-2 production or activity. NO synthesis in endothelial cells is stimulated by various angiogenic factors including FGF-2 [52], VEGF [53, 54] and substance P [55]. Endothelial cell proliferation induced by the specific endothelial growth factor VEGF is NO-dependent and the angiogenic effect of PGE and VEGF in the rabbit cornea is modulated by NO [37, 45]. In addition, the release of angiogenic factors from macrophages appears to be stimulated by NO, suggesting that pro-angiogenic effects of this molecule *in vivo* may be in part independent of direct effects on endothelial cells [56]. NO is also reported to play a pro-angiogenic role in wound healing [57] and in tumours thereby enhancing tumour growth [58]. Exposure of endothelial cells to a hypoxic environment, a powerful stimulus to angiogenesis, also results in the induction of iNOS [59].

In keeping with a proangiogenic role for NO, the angiostatic factor TGFβ-1 inhibits iNOS in endothelial cells [60, 61, 62]. This cytokine could thus counteract the enhanced expression of iNOS and NO synthesis in endothelial cells induced by other angiogenic mediators such as TNFα or IL-1, which may use NO as an effector molecule [60, 63]. However, TGFβ also stimulates the production of the constitutively expressed endothelial cell NOS (eNOS) [64]. This finding in combination with the other available data on the effect of modulating NO on angiogenesis raises the possibility that the effect of NO may depend on the source of NO. Thus exogenous NO may exert different effects on endothelial cell function from endothelial cell derived NO, and further, there may be differences in response to NO from iNOS and eNOS from within the endothelial cells. However, data on the effects of growth factors such as TGFβ on parameters of angiogenesis *in vitro* are not easily translated into effects on the angiogenic process *in vivo*. TGFβ-1 has been shown to have both pro- and anti-angiogenic effects in the CAM [65], with low concentrations inducing a pro-angiogenic response and higher levels causing inhibition. It has also been shown to be a powerful angiogenic stimulus in the rabbit corneal model [66]. However the pro-angiogenic effects of TGFβ are generally thought to be largely due to the recruitment of inflammatory cells, particularly macrophages, and it is these which produce the angiogenic mediators, thus masking the true angiostatic effects of this cytokine which are evident *in vitro*. This difficulty in dissecting the effect of mediators on angiogenesis *in vivo* free from their effects on other systems and subsequent secondary effects on angiogenesis is one of the biggest hurdles to overcome in the field. Due to the complexity of the inflammatory response, this problem is magnified manyfold and requires careful consideration when ascribing roles for mediators in the control of angiogenesis.

In contrast to the above studies suggesting NO is angiogenic, Maragoudakis and co-workers have published a series of reports illustrating that NO may inhibit angiogenesis. They have shown that the NO donor sodium nitroprusside (SNP) inhibited angiogenesis in the CAM assay [67], and antagonised the angiogenic effects of thrombin and PMA. Similarly, tube formation in the *in vitro* matrigel assay was inhibited by SNP. They further demonstrated that the NOS inhibitors L-N^G-monomethyl arginine (L-NMMA) and N$^\omega$-nitro-L-arginine methyl ester (L-NAME) potentiated blood vessel formation in the CAM assay in an L-arginine reversible fashion. Later studies extended these findings to other NO donors (isosorbide 5-mononitrate and isosorbide dinitrate) which exhibited similar angiostatic effects on the CAM, and also inhibited the growth and metastasis of i.p. lewis lung carcinomas in mice [68]. The same authors reported that the angiostatic effects of IL-2 in the CAM were mediated by NO and could be blocked by the inhibition of NOS activity [69].

These contrary effects may be related to the level of NO, with cytotoxic effects at high concentration and angiogenic effects at low concentration. From the literature and in common with many mediators, it is likely that the effects of NO on the angiogenic response will be concentration, site and context-sensitive.

Prostaglandins/nitric oxide and angiogenesis

One interesting aspect to emerge from recent research is the signalling difference between FGF-2 and VEGF in promoting angiogenesis. From the studies quoted here the evidence is quite strong that the effects of VEGF in angiogenesis are mediated (at least initially) through NO, whereas the angiogenic effects of FGF-2 in the CAM assay do not require this molecule. Elegant studies by Ziche and her group have utilized the rabbit corneal model to highlight the differences. These authors have shown that the NOS inhibitor L-NAME inhibits VEGF-transfected MCF-7 cell capillary formation, but not when transfected with FGF-2 [44]. In addition, the angiogenic effects of TNFα and PAF in the CAM are inhibited by NO synthesis inhibition, but that of FGF-2 is not [70]. One suggestion is that FGF-2 lies downstream from NO in the generation of the angiogenic response from VEGF and other angiogenic factors which possess vasodilatory activity and have a dependency on NO production [49, 51]. The angiogenic activity of FGF-2 does however seem to require prostaglandin synthesis [71] and FGF-1 and -2 induce endothelial cell expression of COX-2 [72, 73]. This may explain how endogenous NO production by endothelial cells stimulates eicosanoid synthesis [74]; NO induces FGF-2 activity which in turn induces COX-2 and thus leads to prostaglandin synthesis. However, the application of NO-donors appears to give conflicting results [75, 76]. Thus NO from the NO donor s-adenosyl-N-acetyl-D, L-cetylpenicillamine (SNAP) has been shown to antagonise endothelial cell proliferation, scattering of cells and alterations in cell morphology induced by FGF-2 [77]. The interplay between these two key mediator systems in controlling outcome of the angiogenic response is the subject of much research and still remains to be defined.

Heme oxygenase, carbon monoxide and angiogenesis

In comparison to COX and NOS, the role of HO in angiogenesis has received almost no attention. As with tissue expression of VEGF [78], HO-1 has been demonstrated to increase in hypoxic conditions via the activation of hypoxia-inducible factor (HIF-1) [79]. One would predict from this that HO-1 may be expressed in endothelial cells in situations where an angiogenic response is a result of physiological or pathological tissue expansion in which hypoxia is often a key driving force, such as in the rheumatoid synovium. Abraham and colleagues [80] have demonstrated that transfection of human HO-1 into rabbit coronary endothelial cells could significantly protect the cells from cytotoxic, oxidative stress induced by free heme/hemoglobin. Recently, this overexpression has been shown to result in a 45% increase in basal endothelial cell proliferation in comparison to controls [81]. This was accompanied by a doubling in capillary formation *in vitro* in response to application of FGF-2. This suggests therefore that oxidative stresses, in particular hypoxia, may lead to an increased endothelial expression of HO-1, the protection of those cells from a noxious environment and a facilitation of angiogenesis. Hypoxia has also been shown to lead to a down-regulation of eNOS [82] by reducing transcription and decreasing mRNA stability. This supports the possibility that eNOS may play an inhibitory control in angiogenesis in keeping with its induction by TGFβ [64]. Heme oxygenase induction may also serve to protect the cell against the damaging effects of NO. Thus, HO-1 activity in endothelial cells is increased when these cells are cultured in the presence of NO donors [83] whilst thiol based NO scavengers such as N-acetyl cysteine can reduce the activity of HO-1 by a glutathione dependent mechanism [84]. Similarly, modification of HO-1 activity feeds back onto the NOS synthetic systems in endothelial cells. Zinc protoporphyrin, which is a non-specific inhibitor of HO-1 and HO-2, has been shown to elevate the levels of eNOS mRNA [85] in rat endothelial cells. As mentioned previously, this may lead to a predisposition towards inhibiting angiogenesis, either directly by changing endothelial function or indirectly by adversely affecting endothelial cell survival. Indeed initial studies in our department have shown that the use of an HO inhibitor, tin protoporphyrin, inhibits angiogenesis in an *in vivo* model of chronic inflammation, the murine chronic granulomatous tissue air pouch. *In vivo*, there may also be a role similar to NO for CO production by HO-1 acting as a vasodilator and enhancing vascular permeability [26]. This activity may help promote the angiogenic response by increasing fibrin deposition, leading to the generation of proangiogenic fibrin degradation products [86]. However, despite the evidence for this potential pathway, induction of HO-1 has been shown to inhibit acute inflammatory responses and inihibition of this enzyme using tin protoporphyrin leads to an exacerbation of exudation and cell influx [27, 87]. The contribution of CO to such mechanisms may therefore be minor in comparison to other mediators and mechanisms such that it may be very difficult to demonstrate, particularly *in vivo* in inflammation. Indeed, its effect like NO may be concentration and environment dependent.

Further research in this newly opened area is needed. However, and in particular *in vivo*, more specific HO-1 and HO-2 agonists and antagonists, other than the porphyrin based compounds currently available, are required to elucidate the role of HO and its metabolites in the angiogenic response.

Anti-inflammatory drugs and the angiogenic response

There are relatively few quantitative studies on the effects of drugs on angiogenesis in models of inflammation, mainly due to the difficulty of the assay systems. These vary from what may be considered 'pure' systems where attempts are made to limit the variety of inherent stimuli to a specific applied factor such as the matrigel assay [88], the disc assay [89], the rat mesentery assay [90], and embryonic angiogenesis in the CAM assay [91] to full scale inflammatory lesions such as the murine chronic granulomatous tissue air pouch [92, 93], the cotton pellet [6, 7, 94], delayed type hypersensitivity (DTH) [6] or sponge granulomas [95]. All *in vivo* models of angiogenesis may be complicated by the influence of endogenous inflammatory factors as mentioned previously, and drugs may therefore be active in modulating angiogenesis under these circumstances by affecting the inflammatory response. This point needs careful consideration when examining the results obtained. The method of choice developed in our laboratories for the study of angiogenesis in inflammatory models involves the formation of a vascular cast by the injection of carmine red dye in gelatin [93]. To avoid the problems of vascular dilatation and enhanced vascular permeability inherent in any inflammatory model, animals are maximally vasodilated by warming to $40°C$ prior to injection of casting material. The dye chosen is particulate in nature and has been demonstrated to be retained within the vasculature even in areas of increased vascular permeability. Thus a measurement of vascularity can be derived by expressing the dye content of tissue as a function of the mass of that tissue, independent of the vasoactive effects of drugs.

Most of the available data is from studies using non-steroidal anti-inflammatory drugs to influence blood vessel growth. Indomethacin inhibits corneal neovascularization and FGF-2-induced angiogenesis, both of which can be reversed by prostaglandin replacement [96, 71]. The super-induction of sponge angiogenesis by either FGF-2 or epithelial growth factor, which also induces COX-2 expression, is inhibited by the selective COX-2 inhibitor NS398 [97]. However, in a system which is not stimulated by an exogenous growth factor, the murine chronic granulomatous tissue air pouch, inhibitors which are more selective *in vitro* for the constitutive isoform COX-1 [98] such as indomethacin, piroxicam, ibuprofen and aspirin have mixed effects (Fig. 2). In this study, piroxicam was ineffective, but indomethacin and ibuprofen appeared to give a reduction in vascularity, but this was largely due to a significant increase in granuloma mass of these animals [98]. The dual COX1/2 inhibitor diclofenac given orally at high doses was also without effect, whereas selective inhibition of COX-2 in this system, using nimesulide, (which is reported to have several other actions [99] in addition to COX-2 selectivity [100]), resulted in an increase in granulomatous tissue vascular density. However, along with increased tissue vascularity, nimesulide treatment resulted in increased levels of PGE_2. This was possibly synthesised by COX-1 [101] due to the enhanced free radical scavenging properties of nimesulide [102, 103] which as a function of a pKa > 7 would encourage conversion of PGG to PGH by the constitutive isoform [104]. Examples of a new generation of highly selective COX-2 inhibitors, L-745.337 and NS398 appeared to have no effect on vascularity in this system. Taken together it appears from these data, that at the time point studied, in this model, there was no consistent pattern to inhibition of inflammatory angiogenesis by COX-1, COX-2 or dual inhibitors.

However, the local administration of the dual COX 1/2 inhibitor diclofenac, either directly into the air pouch or applied topically in hyaluronan as a drug delivery agent resulted in inhibition of angiogenesis (Fig. 3) in both cases [105, 106]. In addition, it appeared that neovascular regression could be induced by topical administration (Fig. 4, [105, 106]) in a fashion similar to that seen with angiostatic steroid therapy [4, 107]. The item to note with the delivery system is that hyaluronan sustains and controls the release of diclofenac to local sites from the epidermis, rather than enhancing permeation through the skin for systemic absorption [108, 109]. Prostaglandin synthesis inhibition under these circumstances was greatly enhanced by the local delivery (Tab. 1) when compared to oral administration [110]. The effects of this locally administered diclofenac almost totally inhibited prostaglandin synthesis within the granulomatous tissue in comparison to oral administration (88% compared to 65%) and inhibited angiogenesis. In addition to effects on prostaglandin synthesis, results from our laboratory also indicate that diclofenac may inhibit iNOS activity. Diclofenac inhibits interferon-γ and lipopolysaccharide-induced NO synthesis by J774 murine macrophages, and dramatically reduces iNOS expression [111]. In the murine chronic granulomatous air

Figure 2
The development of the vasculature in inflammation in the presence of inhibitors of COX-1 (aspirin, piroxicam, indomethacin, ibuprofen), COX-1 and COX-2 (diclofenac), and COX-2 (nimesulide, L745, 337, NS-398). Mice with air pouches stimulated with Freund's complete adjuvant and 0.1% croton oil were treated orally, and at termination (6 day granulomatous tissue) carmine/gelatin vascular casts were made of the granulomatous tissue vasculature. The tissues were dried, weighed and macerated and the carmine assayed and the vascular density expressed as mg carmine/mg dry mass tissue. * = p < 0.05, ** = p < 0.01 vs vehicle control.

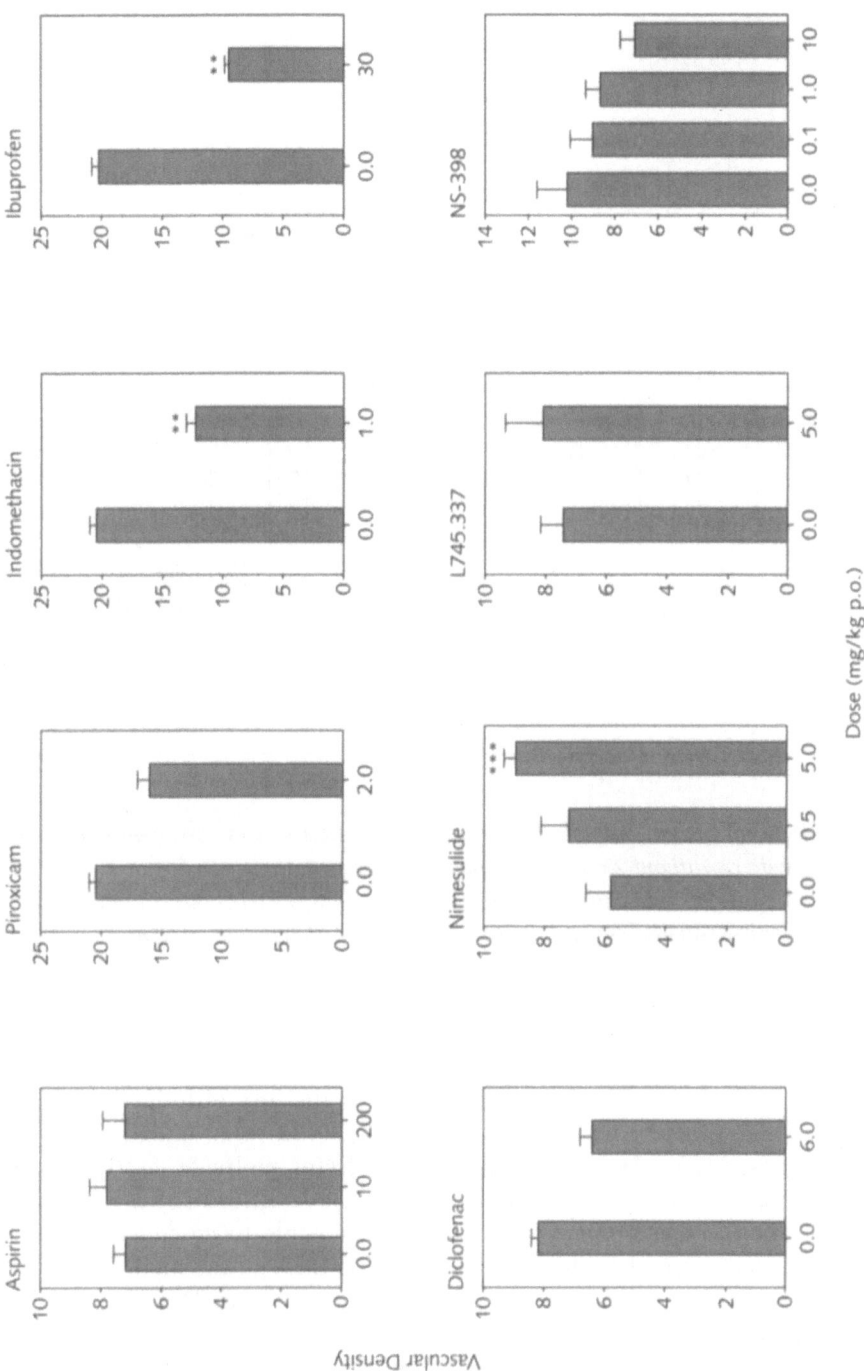

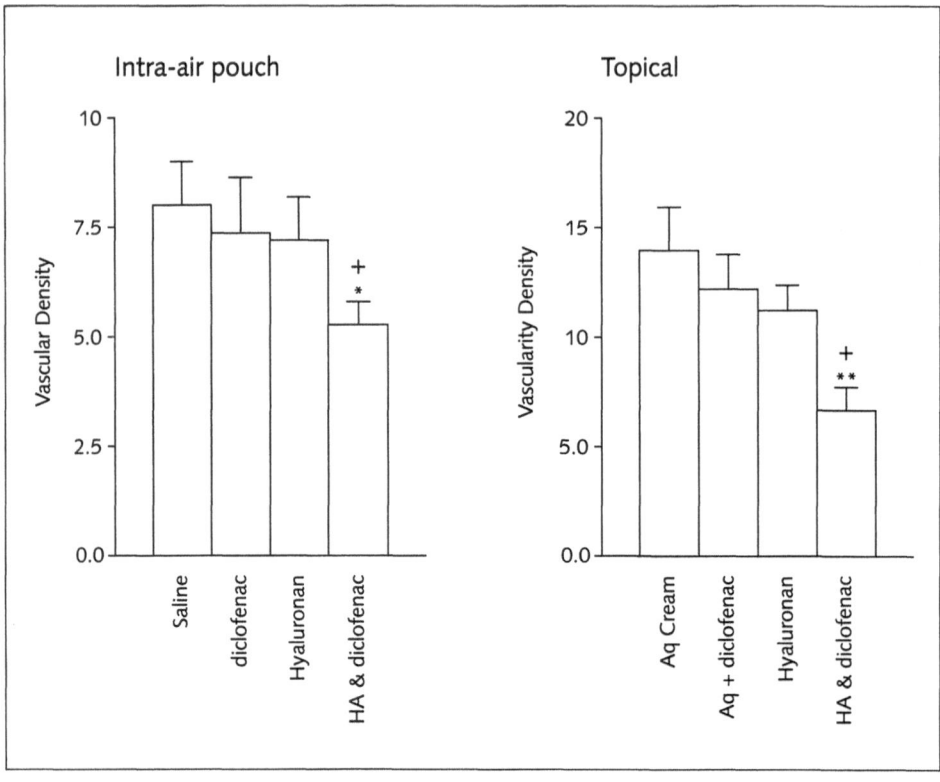

Figure 3
The development of the vasculature in inflammation after 6 days treatment with 6 mg/kg diclofenac with or without 2.5% hyaluronan, method as described in Figure 1. The animals were dosed with diclofenac in saline or in 2.5% hyaluronan (HA) directly intra pouch, or topically in either aqueous cream B.P. (Aq Cream) or 2.5% hyaluronan (HA). + = p < 0.05 vs aqueous cream with diclofenac, * = p < 0.05 and ** = p < 0.01 vs aqueous cream control.

pouch, we have shown VEGF is expressed within the first 24 h during the exudative phase and anti-VEGF therapy is profoundly effective at inhibiting the angiogenic response [112]. As NO is a proposed effector molecule in VEGF-induced angiogenesis, under the circumstances of high levels of diclofenac achieved on topical application in hyaluronan, it would be reasonable to speculate that in this model, diclofenac may be inhibiting angiogenesis through the modulation of NO synthesis in addition to its effects on prostaglandin synthesis. This may explain the lack of effect with oral administration of non-steroidal antiinflammatory drugs, inhibition of both enzyme systems being required for full efficacy.

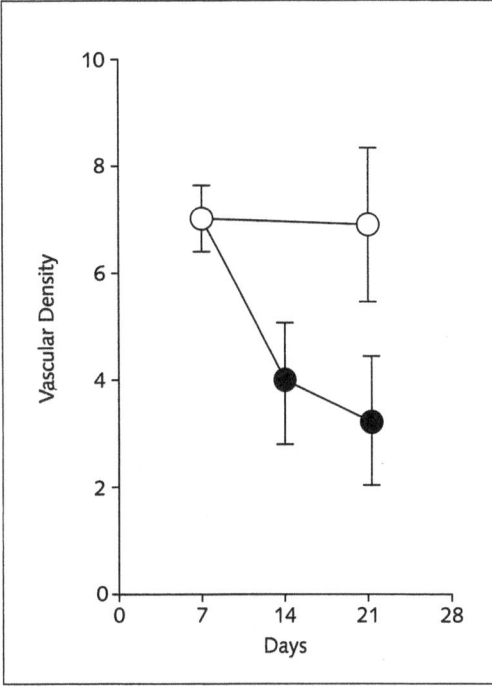

Figure 4
Neovascular regression induced by the topical administration of 6 mg/kg diclofenac in 2.5% hyaluronan (closed circles) applied daily from day 7 to murine chronic granulomatous air pouches (see Fig. 1 for method).

Table 1 - The inhibition of PGE_2 synthesis within murine chronic granulomatous tissue after the oral administration, or topical application in 2.5% hyaluronan, of 6 mg/kg diclofenac to the depilated skin overlying it. The granulomatous tissues contained 73% more PGE_2 than the overlying skin (323.2 ± 84.8, $n = 8$, compared to 187.3 ± 21.4 pg PGE_2/mg, $n = 7$).

Treatment	PGE_2 (pg/mg)	% inhibition
Control	323.2 ± 84.8	
6 mg/kg diclofenac p.o.	114.3 ± 35.1***	65
6 mg/kg diclofenac topical in 2.5% hyaluronan	39.1 ± 13.0***	88

The lack of specificity of the iNOS inhibitors has proven to be a problem in elucidating angiogenic mechanisms. Endothelial cell NOS is a regulator of peripheral vascular tone and administration of selective iNOS inhibitors such L-N[6]-(iminoethyl)-lysine (L-NIL), aminoguanidine and S-(2-aminoethyl)-isothiourea none the less reduce microvascular blood flow, which can be reversed by the addition of

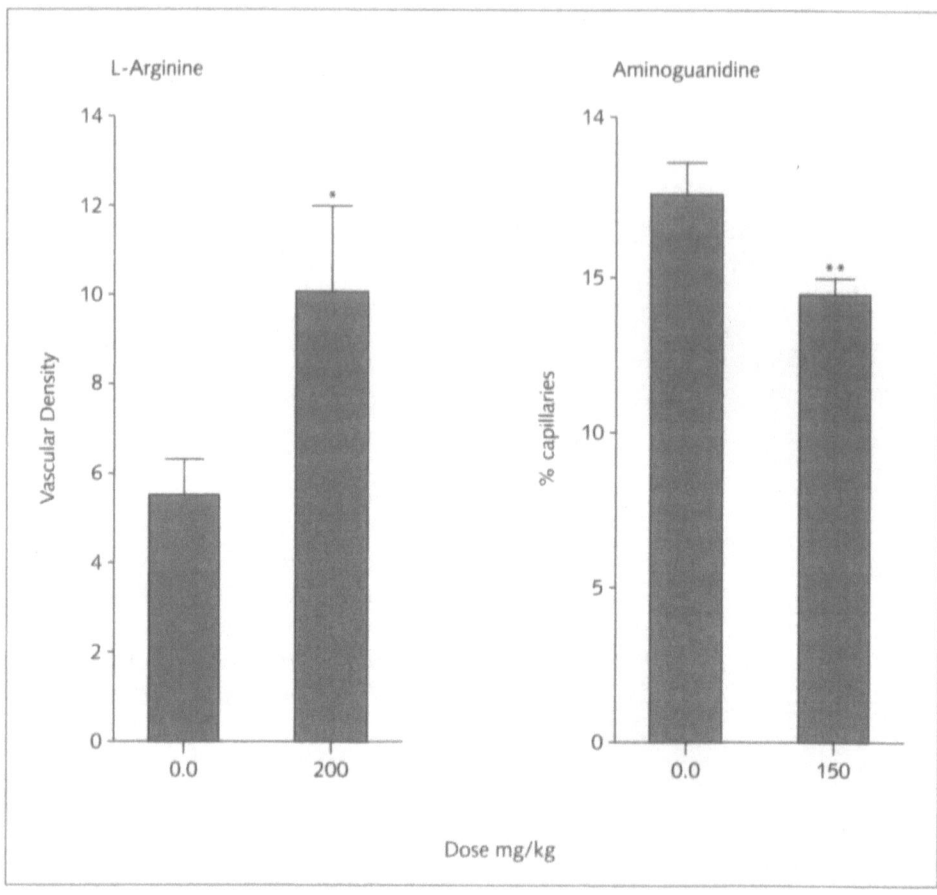

Figure 5
*Nitric oxide stimulation and inhibition in the murine chronic granulomatous air pouch angiogenesis (see Fig. 1 for method). Mice were dosed with the NOS substrate L-arginine and the iNOS inhibitor aminoguanidine in drinking water. % capillaries was calculated histologically as the % capillaries per unit area of random views of 10 µm sections of granulomatous tissues. * = p < 0.05, ** = p < 0.01 vs water controls.*

vasodilators. This may influence the blood flow within inflammatory tissues and interfere with the angiogenic response as well as with the measurement of vascular volume [113–116]. However, from work in our laboratory, we do know that the administration of L-arginine profoundly exaggerates angiogenesis in the murine chronic granulomatous tissue air pouch, already a hyper-vascular tissue (Fig. 5). Furthermore, aminoguanidine (150 mg/kg) administered in drinking water was shown by morphometric analysis to significantly reduce the number of CD31 immunolabelled endothelial cells in comparison to controls. This measurement of vascularity, as opposed to the carmine/gelatin vascular cast is unaffected by peripheral constriction by the NOS inhibitors. It would appear therefore that within this inflammatory locus NO is angiogenic.

Thus in summary, the inducible enzymes COX-2, iNOS and HO-1 may have important roles to play in the mediation of angiogenesis. COX-2, or rather its metabolites PGE_2 and PGI_2, have definite pro-angiogenic actions, and NO produced by iNOS appears to be pro-angiogenic in inflammation as well. The protective role of HO-1 induction in endothelial cells also points to a role supporting the angiogenic response, particularly in hypoxic environments. Further work in our department is examining the role of these enzymes in other pathologies such as tumorigenesis [117] and their role in controlling the angiogenic response as part of disease pathogenesis. The implications of theraputic strategies aimed at targeting these enzyme systems for angiogenesis-dependent processes, both pathological and physiological, are however still largely to be elucidated.

Acknowledgements

The authors gratefully acknowledge the generous financial support of the Arthritis and Rheumatism Council (United Kingdom) and The Joint Research Board of The Special Trustees of St Bartholomew's Hospital.

References

1   Folkman J (1986) How is blood vessel growth regulated in normal and neoplastic tissue. GHA Clowes Memorial Award Lecture. *Cancer Res* 46: 467–473

2   Colville-Nash PR, Scott DL (1992) Angiogenesis and rheumatoid arthritis: pathogenic and therapeutic implications. *Ann Rheum Dis* 51: 919–925

3   Colville-Nash PR, Willoughby DA (1997) Growth Factors in angiogenesis: current interest and therapeutic potential. *Molecular Medicine Today* 3: 14–23

4   Seed MP, Colville-Nash PR, Jackson JR, Winkler J (1998) Angiogenesis in inflammation. In: T-P Fan, R Auerbach (eds) *The New Angiotherapy*, Humana Press, USA

5   Seed MP (1996) Angiogenesis inhibition as a drug target for disease: An update. *Exp Opin Invest Drugs* 5: 1617–1637

6   Dunn CJ, Galinet LA (1991) Angiostatic cortexone-heparin combination treatment suppresses chronic granulomatous inflammation in mice. *Drug Devel Res* 23: 241–248

7   Colville-Nash PR, El-Ghazaly M, Willoughby DA (1993) The use of angiostatic steroids to inhibit cartilage destruction in an *in vivo* model of granuloma mediated cartilage degradation. *Agents Actions* 38: 126–134

8 Josefsson E, Tarkowski A (1997) Suppression of type II collagen-induced arthritis by the endogenous estrogen metabolite 2-methoxyestradiol. *Arth Rheum* 40: 154–163

9 Peacock DJ, Banquerigo ML, Brahn E (1995) A novel angiogenesis inhibitor suppresses rat adjuvant arthritis. *Cell Immunol* 160: 178–184

10 Peacock DJ, Banquerigo ML, Brahn E (1992) Angiogenesis inhibition suppresses collagen arthritis. *J Exp Med* 175: 1135–1138

11 Colville-Nash PR, Seed MP (1993) The current state of angiostatic therapy, with special reference to rheumatoid arthritis. *Curr Opin Invest Drugs* 2: 763–813

12 Colville-Nash PR, Seed MP, Willoughby DA (1992) Antirheumatic drugs and the development of vasculature in murine chronic granulomatous air pouches. *Br J Pharmacol* 107: 421P

13 Petersen HI (1982) Tumour angiogenesis inhibition by prostaglandin synthetase inhibitors. *Anticancer Res* 6: 251–253

14 Petersen H (1983) Effects of prostaglandin synthesis inhibitors on tumour growth and neovascularisation. *Invasion Metastases* 3: 151–159

15 Pober S, Cotran R (1990) Cytokines and endothelial cell biology. *Physiol Rev* 70: 427–451

16 Salgardo A, Boveda JL, Monasterio J, Segura RM, Mourelle M, Gomez-Jiminez J, Peracula R (1994) Inflammatory mediators and their influence on haemostasis. *Haemostasis* 24: 132–138

17 Hosaka S, Shah MR, Barquin N, Haines GK, Koch AE (1995) Expression of fibroblast growth factor and angiogenesis in arthritis. *Pathobiology* 63: 249–256

18 Lupia E, Montrucchio G, Battaglia E, Modena V, Camussi G (1996) Role of tumour necrosis factor-α and platelet activating factor in neoangiogenesis induced by synovial fluids of patients with rheumatoid arthritis. *Eur J Immunol* 26: 1690–1694

19 Norrby K (1997) Interleukin-1α and *de novo* mammalian angiogenesis. *Microvasc Res* 54: 58–64

20 Jackson JR, Minton JA, Ho ML, Wei N, Winkler JD (1997) Expression of vascular endothelial growth factor in synovial fibroblasts is induced by hypoxia and interleukin-1β. *J Rheumatol* 24: 1253–1259

21 Byrd V, Zhao XM, McKeehan WL, Miller GG, Thomas JW (1996) Expression and functional expansion of growth factor receptor T cells in rheumatoid synovium and peripheral blood of patients with rheumatoid arthritis. *Arthritis Rheum* 39: 914–922

22 Mapp, PI and Blake, DR (1995) Neuropeptides and the synovium. In: B Henderson, JCW Edwards, ER Pettipher (eds) *Mechanisms and Models in Rheumatoid Arthritis*. Academic Press, 317–333

23 Dusting GJ, Moncada S, Vane JR (1978) Vascular actions of arachidonic acid and its metabolites in the perfused mesenteric and femoral beds of the dog. *Eur J Pharmacol* 49: 65–72

24 Moncada S, Palmer RMJ, Higgs EA (1991) Nitric oxide, Physiology, pathophysiology and pharmacology. *Pharmacol Rev* 43: 109–142

25 Maines MD (1988) Heme oxygenase: function, multiplicity, regulatory mechanisms and clinical applications. *Faseb J* 2: 2557–2568

26 Furchgott RF, Jthianandan D (1991) Endothelium dependent and endothelium independent vasodilation involving cyclic-GMP-relaxation induced by nitric oxide, carbon monoxide, and light. *Blood Vessels* 28: 52–61

27 Willis D, Moore AR, Frederick R, Willoughby DA (1996) Heme oxygenase: A novel target for the modulation of the inflammatory response. *Nature Med* 2: 87–90

28 Tomlinson A, Appleton I, Moore AR, Gilroy DW, Willis D, Mitchell JA Willoughby DA (1994) Cyclo-oxygenase, and nitric oxide synthase isoforms in rat carrageenin-induced pleurisy. *Br J Pharm* 113: 693–698

29 Appleton I, Tomlinson A, Mitchell JA, Willoughby DA (1995) Distribution of cyclooxygenase isoforms in murine chronic granulomatous inflammation. Implications for future anti-inflammatory therapy. *J Pathol* 176: 413–420.

30 Crofford LJ, Wilder RL, Ristimaki AP, Sano H, Remmers EF, Epps HR, Hla T (1994) Cyclooxygenase-1 and -2 expression in rheumatoid synovial tissues. Effects of interleukin-1β, phorbol ester and corticosteroids. *J Clin Invest* 93: 1095–1101

31 Tomlinson A, Appleton I, Moore AR, Willis D, Gilroy D, Willoughby DA (1994) Localisation of isoforms of cyclooxygenase (COX) and nitric oxide synthase (NOS) in models of acute and chronic inflammation. Proceedings of 9th Int. Conf. Prostaglandins and Related Compounds, p 37

32 Grabowski PS, Wright PK, Van't Hof RJ, Helrich MH, Ohshima H, Ralston SH (1997) Immunolocalization of inducible nitric oxide synthase in synovium and cartilage in rheumatoid arthritis and osteoarthritis. *Br J Rheumatol* 36: 651–655

33 Sakurai H, Kohsaka H, Liu MF, Higashiyama H, Hirata Y, Kanno K, Saito I, Miyasaka N (1995) Nitric oxide production and inducible nitric oxide synthase expression inflammatory arthritides. *J Clin Invest* 96: 2357–2363

34 Groszanovic Z, Gossrau R (1996) Expression of heme oxygenase-2 (HO-2)-like immunoreactivity in the rat. *Acta Histochem* 98: 203–214

35 Zakhary R, Gaine SP, Dinerman JL, Ruat M, Flavahan NA, Snyder SH (1996) Heme-oxygenase-2: endothelial and neuronal localization and role in endothelium dependent relaxation. *Proc Nat Acad Sci* 93: 795–798

36 Ziche M, Jones J, Gullino PM (1982) Role of prostaglandin E$_1$ and copper in angiogenesis. *J Natl Cancer Inst* 69: 475–482

37 Ziche M, Morbidelli L, Parenti A, Ledda F (1995) Nitric Oxide modulates angiogenesis elicited by prostaglandin E(1) in rabbit cornea. *Adv Prostaglandin Thromboxane Res* 23: 495–497

38 Form DM, Auerbach R (1983) PGE$_2$ and angiogenesis. *Proc Soc Exp Med* 172: 214–218

39 Diazflores L, Gutierrez R, Valladares F, Varela H, Perez M (1994) Intense vascular sprouting from rat femoral vein induced by prostaglandins-E1 and E2. *Anatomical Record* 238: 68–76

40 Oktsu A, Fujii K, Kurozumi S (1988) Induction of angiogenic response by chemically

stable prostacyclin analogues. *Prostaglandins Leukotrienes Essential Fatty Acids* 33: 35–39

41   Yamamoto T, Horikawa N, Komuro Y, Hara Y (1996) Effect of topical application of a stable prostacyclin analogue, SM-10902 on wound healing in diabetic mice. *Eur J Pharmacol* 302: 53–60

42   Lewis GP (1986) Products derived from arachidonic acid. In: GP Lewis (ed) *Mediators of inflammation*. Wright Publ, Bristol, 44–58

43   Spencer-Green G, Caulkins KM (1993) Augmentation of interleukin-1 induced prostacyclin production by endothelial cell growth factor: Implications for chronic synovitis. *Prostaglandins* 45: 439–445

44   Pankonin G, Teuscher E (1991) Stimulation of endothelial cell migration by thrombin. *Biomed Biochim Acta* 50: 1073–1078

45.   Ziche M, Morbidelli L, Choudhuri R, Zhang H-T, Donnini S, Granger, H, Bicknell (1997) Nitric oxide synthase lies downstream from vascular endothelial growth factor-induced but not basic fibroblast growth factor -induced angiogenesis. *J Clin Invest* 99: 2625–2634

46   Nagashima M, Yoshino S, Ishiwata T, Asano G (1995) Role of vascular endothelial growth factor in angiogenesis of rheumatoid arthritis. *J Rheumatol* 22: 1624–1630

47   Fava RA, Olsen NJ, Spencer-Green G, Yeo KT, Berse B, Jackman RW, Senger DR, Dvorak HF, Brown LF (1994) Vascular permeability factor/endothelial growth factor (VPR/VEGF): accumulation and expression in human synovial fluids and rheumatoid synovial tissue. *J Exp Med* 180: 341–346

48   Koch AE, Harlow LA, Haines GK, Amento EP, Unemori EN, Wong WL, Pope RM, Ferrarra N (1994) Vascular endothelial growth factor. A cytokine modulating endothelial function in rheumatoid arthritis. *Immunol* 152: 4149–4156

49   Ben-AV P, Crofford LJ, Wilder RL, Hla T (1995) Induction of vascular endothelial growth factor expression in synovial fibroblasts by prostaglandin E and interleukin-1: a potential mechanism for inflammatory angiogenesis. *FEBS Lett* 372: 83–7

50   Morbidelli L, Chang CH, Douglas JG, Granger HJ, Ledda F, Ziche M (1996) Nitric oxide mediates mitogenic effect of VEGF on coronary venular endothelium. *Am J Physiol* 39: H411–H415

51   Ziche M, Parenti A, Ledda F, Dell'Era P, Granger HJ, Maggi CA, Presta M (1997) Nitric oxide promotes proliferation and plasminogen activator production by coronary venular endothelium through endogenous bFGF. *Circ Res* 80(6): 845–852

52   Tiefenbacher CP, Chilian WM (1997) Basic fibroblast growth factor and heparin influence coronary arteriolar tone by causing endothelium dependent dilation. *Cardiovasc Res* 34: 411–417

53   Fujii E, Ire K, Ohba K, Ogawa A, Yoshioka T, Yamakawa M (1997) Role of nitric oxide, prostaglandins and tyrosine kinase in vascular endothelial growth factor-induced increase in vascular permeability in mouse skin. *Naunyn Schmiedebergs Arch Pharmacol* 356: 475–480

54   van der Zee R, Murohara T, Luo ZY, Zollmann F, Passeri J, Lekutat C, Isner JM (1997)

Vascular endothelial growth factor vascular permeability factor augments nitric oxide release from quiescent rabbit and human vascular endothelium. *Circulation* 95: 1030–1037

55 Ziche M, Morbidelli L, Masini E, Amerini S, Granger HJ, Maggi CA, Geppeti P, Ledda F (1994) Nitric oxide mediates angiogenesis *in vivo* and endothelial cell growth and migration *in vitro* promoted by substance-P. *J Clin Invest* 94: 2036–2044

56 Leibovitch SJ, Polverini PJ, Fong TW, Harlow LA, Koch AE (1994) Production of angiogenic activity by human monocytes requires an l-arginine nitric oxide synthase-dependent effector. *Proc Natl Acad Sci USA* 91: 4190–4194

57 Konturek SJ, Brzozowski T, Majika J, Purko-Polonczyk J, Stachuro J (1993) Inhibition of nitric oxide synthase delays healing of chronic gastric ulcers. *Eur J Pharmacol* 239: 215–217

58 Jenkins DC, Charles IG, Thomsen LL, Moss DW, Holmes LA, Bayliss SA, Rhodes P, Westmore K, Emson PC, Moncada S (1995) Roles of nitric oxide in tumour growth. *Proc Natl Acad Sci USA* 92: 4392–4396

59 Mellilo G, Musso T, Sica A, Taylor LS, Cox GW, Varesio LA (1995) A Hypoxia-responsive element mediates a novel pathway of activation of the inducible nitric oxide promoter. *J Exp Med* 182: 1683–1693

60 Schneemann M, Schoeden G, Frei K, Schaffner A (1993) Immunovascular communication: Activation and deactivation of murine endothelial cell nitric oxide synthase by cytokines. *Immunol Lett* 35: 159–162

61 Murata J, Corradin SB, Felley-Bosco E, Juillerat-Jeanneret L (1995) Involvement of a transforming growth factor beta-like molecule in tumour cell derived inhibition of nitric oxide synthesis in cerebral endothelial cells. *Int J Cancer* 62: 743–748

62 Kanno K, Hirata Y, Imai T, Iwashina M, Marumo F (1994) Regulation of inducible nitric oxide synthase gene by interleukin-1 beta in rat vascular endothelial cells. *Am J Physiol* 267: H2318–2324

63 Ungureanu-Longrois D, Ballingand JL, Simmons WW, Okada I, Kobzik L, Lowenstein CJ, Kunkel SL, Michel T, Kelly RA, Smith TW (1995) Induction of nitric oxide synthase activity by cytokines in ventricular myocytes is necessary but not sufficient to decrease contractile responsiveness to beta adrenergic agonists. *Circ Res* 77: 494–502

64 Inoue N, Venema RC, Sayegh HS, Ohara Y, Murphy TJ, Harrison DG (1995) Molecular regulation of the bovine endothelial cell nitric oxide synthase by transforming growth factor beta1. *Arterioscler Thromb Vasc Biol* 15: 1255–1261

65 Yang EY, Moses HL (1990) Transforming growth factor beta 1-induced changes in cell migration, proliferation and angiogenesis in the chicken chorioallantoic membrane. *J Cell Biol* 111: 731–741

66 Fiegel VD, Knighton DR (1988) Transforming growth factor beta causes indirect angiogenesis by recruiting monocytes. *FASEB J* 137: 295–302.

67 Pipili-Synetos E, Sakkoula E, Haralabopoulos G, Andriolopoulou, Peristeris P, Maragoudakis ME (1994) Evidence that nitric oxide is an endogenous antiangiogenic mediator. *Br J Pharmacol* 111: 894–902

68 Pipili-Synetos E, Papageorgiou A, Sotiropolou G, Fotsis T, Karaliulakis G, Maragoudakis ME (1995) Inhibition of angiogenesis, tumour growth and metastasis by the NO-releasing vasodilators, isosorbide mononitrate and dinitrate. *Br J Pharmacol* 116: 1829–1834

69 Sakkoula E, Pipili-Synetos E, Maragoudakis ME (1997) Involvement of nitric oxide in the inhibition of angiogenesis by interleukin-2. *Br J Pharmacol* 122: 793–795

70 Montrucchio G, Lupia E, DeMartino A, Battaglia E, Arese M, Tizzani A, Bussolino F, Camussi G (1997) Nitric oxide mediates angiogenesis induced *in vivo* by platelet activating factor and by tumour necrosis factor-α. *Am J Pathol* 151: 557–563

71 Spisni E, Manica F, Tomasi V (1992) Involvement of prostanoids in the regulation of angiogenesis by polypeptide growth factors. *Prostaglandins & Essential Fatty Acids* 47: 111–115

72 Moatter TM, Gerritsen ME (1994) Fibroblast growth factor upregulates PGG/H synthase in rabbit microvascular endothelial cells by a glucocorticoid independent mechanism. *J Cell Physiol* 151: 571–578

73 Moatter TM, Gerritsen ME (1994) Acidic fibroblast growth factor induction of cyclooxygenase-2 in rabbit cardiac muscle microvessel endothelial cells: mediation by protein kinase C. *Microcirculation* 1: 79–88

74 Davidge ST, Baker PN, Laughlin MK, Roberts JM (1995) Nitric oxide produced by endothelial cells increases production of eicosanoids through activation of prostaglandin H synthase. *Circ Res* 77: 274–283

75 DeCatarina R, Dorso C, Tack-Goldman K, Weksler B (1985) Nitrates and endothelial prostacyclin production. *Circulation* 71: 176–182

76 Levin R, Jaffe E, Weksler B, Tack-Goldman K (1981) Nitroglycerin stimulates synthesis of prostacyclin by cultured endothelial cells. *J Clin Invest* 67: 762–769

77 Chaudhury AR, Frischer H, Malik AB (1996) Inhibition of endothelial cell proliferation and bFGF-induced phenotypic modulation by nitric oxide. *J Cell Biochem* 63: 125–134

78 Liu Y, Cox SR, Morita T, Kourembanas S (1991) Hypoxia regulates VEGF gene expression in endothelial cells: Identification of a 5' enhancer. *Circ Res* 77: 635–643

79 Lee PJ, Jiang B-H, Chin BY, Iyer NV, Alam J, Semenza GL, Choi AMK (1997) Hypoxia inducible factor mediates transcriptional activation of the heme oxygenase gene in response to hypoxia. *J Biol Chem* 272(9): 5375–5381

80 Abraham NG, Lavrosky Y, Schwartzman ML, Stoltz RA, Levere RD, Gerritsen ME, Shibahara S, Kappas A (1995) Transfection of the human heme oxygenase gene into rabbit coronary microvessel endothelial cells: protective effect against heme and hemoglobin toxicity. *Proc Natl Acad Sci USA* 92: 6798–6802

81 Deramaudt BM, Braunstein S, Remy P, Abraham NG (1998) Gene transfer of human heme oxygenase into coronary endothelial cells potentially promotes angiogenesis. *J Cell Biochem* 68(1): 121–127

82 Kourembanas S, Morita T, Liu Y, Christou H (1997) Mechanisms by which oxygen regulates gene expression and cell-cell interaction in the vasculature. *Kidney Int* 51: 438–443

83   Yee EL, Pitt BR, Billiar TR, Kim YM (1996) Effect of nitric oxide on heme- metabolism in pulmonary artery endothelial cells. *Am J Physiol* 271: L512–518

84   Foresti R, Clarke JE, Green CJ, Motterlini R (1997) Thiol compounds interact with nitric oxide in regulating heme oxygenase-1 induction in endothelial cells. Involvement of superoxide and peroxynitrite anions. *J Biol Chem* 272: 18411–18417

85   Seki T, Naruse M, Yoshimoto T, Tanabe A, Imaki T, Hagiwara H, Hirose S, Demura H (1997) Interrelation between nitric oxide synthase and heme oxygenase in rat endothelial cells. *Eur J Pharmacol* 331: 87–91

86   Thompson WD, Harvey JA, Kazmi MA, Stout AJ (1991) Fibrinolysis and angiogenesis in wound healing. *J Pathol* 165: 311–318

87   Willis. D. (1995) Expression and modulatory effects of heme oxygenase in acute inflammation in the rat. *Inflammation Res* 44(S2): S218–S220

88   Kibbey MC, Grant DS, Kleinman UK (1992) Role of SIKVAV site of laminin in promotion of angiogenesis and tumour growth – An *in vivo* matrigel model. *J Natl Cancer Inst* 84: 1663–1638.

89   Kowalski J, Kwan HH, Prionas SD, Allison AC, Fajardo LF (1992) Characterization of the disc angiogenesis system. *Exp Mol Pathol* 56: 1–19

90   Jakobsson AE, Norrby K, Ericsson LE (1994) A morphometric method to evaluate angiogenesis kinetics in the rat mesentry. *Int J Exp Pathol* 25: 219–224

91   Ausprunk DH, Knighton DR, Folkman J (1974) Differentiation of vascular endothelium in the chick chorioallantois: a structural and autoradiographic study. *Dev Biol* 38: 237–248

92   Kimura M, Amemiya K, Yamada T, Suzuki J (1986) Quantitative method for measuring adjuvant-induced granuloma angiogenesis in insulin-treated diabetic mice. *J Pharmacobiodyn* 9: 442–446

93   Colville-Nash PR, Alam CAS, Appleton I, Brown JR, Seed MP, Willoughby DA (1995) The Pharmacological modulation of angiogenesis in chronic granulomatous inflammation. *J Pharmacol Exp Ther* 274: 1463–1472

94   Orlandi C, Dunn CJ, Cutshaw LG (1988) Evaluation of angiogenesis in chronic inflammation by laser-Doppler flowmetry. *Clin Sci* 74: 119–121

95   Andrade SP, Fan T-P, Lewis GP (1987) Quantitative studies on angiogenesis in a rat sponge model. *Br J Exp Path* 68: 755–766

96   Frucht J, Zauberman H (1984) Topical indomethacin effect on neovascularization of the cornea and on prostaglandin E$_2$ levels. Br J Ophthalmol 68: 656–659

97   Majima M, Isono M, Ikeda Y, Hayashi I, Hatanaka K, Harada Y, Katsumata O, Yamashina S, Katori M, Yamamoto S (1997) Significant roles of inducible cyclooxygenase (COX)-2 in angiogenesis in rat sponge implants. *Jpn J Pharmacol* 75(2): 105–114

98   Colville-Nash PR, Seed MP, Willoughby DA (1992) Antirheumatic Drugs and the development of vasculature in murine chronic granulomatous air pouches. *Br J Pharmacol* 107: 421P

99   Bevilacqua M, Magni E (1993) Recent contributions to knowledge of the mechanism of action of nimesulide. *Drugs* 46 (S1): 40–47

100 Famaey JP (1997) *In vitro* and *in vivo* pharmacological evidence of selective cyclooxy-genase-2 inhibition by nimesulide: An overview. *Inflamm Res* 46: 437–446

101 Gilroy D, Tomlinson A, Willoughby DA (1998) Differential effects of inhibition of iso-forms of cyclooxygenase (COX-1, COX-2) in chronic inflammation. *Inflamm Res* 47: 79–85

102 Maffei FR, Carini M, Aldini G, Saibene L, Macciocchi A (1993) Antioxidant profile of nimesulide, indomethacin and diclofenac in phosphatidyl choline liposomes (PCL) as membrane model. *Int J Tiss React* XV: 225–234

103 Maffei FR, Carini M, Aldini G, Saibene L, Morelli R (1995) Differential inhibition of superoxide, hydroxyl and peroxyl radicals by nimesulide and its main metabolite 4-hydroxynimesulide. *Arneimittelforschung* 45: 1102–1109

104 Rufer C, Schillinger E, Middleton J, Pons F, Rabeck C, Thierer K, Wintle J, Wolff B, Zsak M, Dukor P (1996) Some aspects of IL-8 pathophysiology. III: Chemokine inter-action with endothelial cells. *J Leukocyte Biol* 31: 3591–3596

105 Alam CAS, Seed MP, Willoughby DA (1995) Angiostasis and vascular regression in chronic granulomatous inflammation induced by diclofenac in combination with Hyaluronan. *J Pharm Pharmacol* 47: 407–411

106 Alam CAS, Seed MP, Willoughby DA (1995) Angiostasis and vascular regression in chronic granulomatous inflammation induced by diclofenac in combination with hyal-uronan (HYAL CT-1101). *Annals Rheum Dis* 54: 757

107 Seed MP, Alam CAS, Willoughby DA (1995) Regression of granulomatous tissue neo-vasculature with angiostatic steroid therapy. *Annals Rheum Dis* 54: 777

108 Brown MB, Marriott C, Martin GP (1995) The effect of hyaluronan on the *in vitro* deposition of diclofenac within the skin. *Int J Tiss React* XVII: 133–140

109 Brown MB, Bennett F, Marriott C, Martin GP (1996) Hyaluronan: a transdermaldrug delivery system. An *in vitro* investigation. *Prog Rheumatol* VI: 59–63

110 Papworth J, Seed MP, Willoughby DA (1996) Resident granulomatous tissue and tumour prostaglandin synthesis inhibition by topical diclofenac in hyaluronan (HYAL EX-0001). *Roy Soc Med Round Table Ser* 45: 54–58

111 Colville-Nash PR, Clarke AE, Sy-Yed, Gilroy DW, Paul-Clark M, Tomlinson A, Willoughby DA (1996) Control of inducible nitric oxide synthase in murine macrophages by diclofenac and hyaluronic acid. *Royal Soc Med Round Table Ser* 45: 127–146

112 Appleton I, Brown NJ, Willis D, Colville-Nash PR, Alam CAS, Brown JR, Willoughby DA (1996) The role of vascular endothelial growth factor in a murine granulomatous tissue air pouch model of angiogenesis. *J Pathol* 180: 90–94

113 Handy RL, Wallacw P, Moore PK (1996) Inhibition of nitric oxide synthase by isoth-ioureas: cardiovascular and anti-nociceptive effects. *Pharmacol Biochem Behav* 55: 179–184

114 Najafipour H, Ferrell WR (1993) Nitric oxide modulates sympathetic vasoconstriction and basal blood flow in normal and healthy inflamed rabbit knee joints. *Exp Physiol* 78 (5): 615–624

115 Ridger VC, Pettipher ER, Bryant CE, Brain SD (1997) Effect of nitric oxide synthase inhibitors aminoguanidine and L-N6-(1-iminoethyl)lysine on zymosan-induced plasma extravsation in rat skin. *J Immunol* 159: 383–390

116 Griffiths MJ, Messent M, MacAllister RJ, Evans TW (1993) Aminoguanidine selectively inhibits iNOS synthesis. *Br J Pharmacol* 110: 963–1604

117 Seed MP, Brown JR, Freemantle CN, Papworth JL, Colville-Nash PR, Willis D, Somerville KW, Asculai S, Willoughby DA (1997) The inhibition of colon-26 adenocarcinoma development and angiogenesis by topical diclofenac in 2.5% hyaluronan. *Cancer Res* 57: 1625–1629

Role of the inducible forms of cyclooxygenase and nitric oxide synthase in inflammatory pain

Sergio H. Ferreira[1], Fernardo Q. Cunha[1] and Stephen Hyslop[2]

[1]Department of Pharmacology, Faculty of Medicine Ribeirão Preto, University of São Paulo, Avenida Bandeirantes, 3900 Ribeirão Preto, São Paulo, Brazil; [2]Department of Pharmacology, Faculty of Medical Science, University of Campinas, Campinas, São Paulo, Brazil

Introduction

Nociceptors and hyperalgesia

The sensitisation of pain receptors is the common denominator in all types of inflammatory pain. C-Polymodal, high threshold receptors or receptors connected by fine myelinated fibers have long been associated with inflammatory hyperalgesia [1, 2]. In recent years, a new "sleeping" nociceptor associated with certain small afferent fibers has been described in deep visceral innervations (colon and bladder) and in joints [3, 4]. Sleeping nociceptors are not active in normal tissues, but are "switched on" during inflammation. This functional upregulation leads to a clinical state known as hyperalgesia. In such a situation, previously unpainful stimuli become painful.

The molecular events associated with hyperalgesia are not yet fully understood. However, there is evidence that an increase in intracellular cAMP and Ca^{2+} concentrations is associated with the functional upregulation of nociceptors. Using a modified rat paw pressure test [5], we have shown that the intraplantar administration of dibutyryl cAMP, Ca^{2+} ionophore or $BaCl_2$ (which increases the concentration of cytosolic free Ca^{2+}) causes hyperalgesia [6]. A similar response is obtained following the administration of prostaglandins or sympathomimetics (noradrenaline or dopamine) known to stimulate neuronal cAMP synthesis. On the other hand, pretreating paws with a calcium channel blocker or with lanthanum (which blocks Ca^{2+} influx) prevents the development of hyperalgesia [6]. The cAMP/Ca^{2+} hypothesis has been substantiated by experimental evidence from other groups using different tests and models [7–10]. The final biochemical events responsible for the functional upregulation of nociceptors following an increase in cytosolic cAMP are not yet understood. The mechanism may involve the activation of a protein kinase A, with subsequent phosphorylation of an ion channel or the modulation of cytosolic structures that control intracellular calcium levels. We have recently proposed that spinal glutamate is involved, and that it causes continuous retrograde sensitisation of primary sensory neurones by opening Ca^{2+} channels [11]. The involvement of tetro-dotoxin-resistant, voltage-dependent Na^+ channels in nociceptor sensitisation has also been suggested [12].

Direct blockade of ongoing hyperalgesia has been observed after local adminis-tration of dibutyryl cGMP or of substances which stimulate neuronal soluble guany-late cyclase (carbachol, acetylcholine or nitric oxide generators) [6–8, 11, 13]. Thus, functional up- or down-regulation of the nociceptors may be dependent on the bal-ance between intracellular cAMP and cGMP levels. Following the discovery of the L-arginine/nitric oxide (NO)/cGMP pathway [14, 15], we demonstrated that NO caused peripheral analgesia by increasing cGMP levels in primary sensory neurones [13, 16].

From a practical point of view, the activation of a nociceptor depends on the co-operation of two distinct groups of stimuli, namely chemical mediators, which induce nociceptor sensitisation, i.e. hyperalgesia, and mechanical, thermal and chemical stimuli that activate sensitised pain receptors. Following such sensitisation, previously ineffective stimuli cause "overt pain" in humans, or a characteristic behaviour in laboratory animals that may be used as an end point in nociceptive tests.

Classic hyperalgesic mediators

The two groups of directly-acting, hyperalgesic mediators which satisfactorily ful-fill most of the experimental and clinical criteria are arachidonic acid/cyclo-oxy-genase products, in particular prostaglandins such as PGE_2, and sympathomimet-ic amines. The ability of prostaglandins to sensitise pain receptors has been demonstrated in humans and in animals using both behavioural and electrophys-iological techniques [17]. Sympathomimetic amines (noradrenaline and dopamine) have been shown to functionally upregulate nociceptors [18]. On the other hand, depleting peripheral sympathomimetic amine stores with guanethi-dine, or treating animals with adrenergic antagonists (β-blockers) or with a dopamine (D1) antagonist (SCH 23390), significantly reduced carrageenan-induced hyperalgesia. These results show that there is a sympathomimetic com-ponent, possibly mediated by D1 receptors, in this model of hyperalgesia [18]. A role for the sympathetic system in hyperalgesia has also been demonstrated by oth-ers [19–22], although the receptors involved seem to vary depending on the model and animal species used. In certain cases, the action of sympathomimetic amines appears to be indirect via cyclooxygenase metabolites, or may even depend on pre-vious inflammatory sensitisation of the tissue. Although in some experimental models of inflammatory pain both prostaglandins and sympathomimetic amines are involved, their relative contribution may well depend on the characteristics of the pathological stimulus.

Cytokines and hyperalgesia

The presence of foreign material in, or injury to, tissue induces an early response that can be envisaged as an alarm reaction in which resident cells seem to play a pivotal role in the development of hyperalgesia and other acute inflammatory events [23].

Although the final peripheral hyperalgesic mediator may be either a prostaglandin or a sympathomimetic amine, these events are secondary to the release of a cascade of cytokines [24, 25]. Using specific antisera for interleukin-1β (IL-1β) and IL-8, as well as cyclooxygenase inhibitors and sympatholytics, we have demonstrated in animal models that these cytokines are responsible for the release of prostaglandins and sympathomimetic amines, respectively. The release of these two cytokines is preceded by the liberation of tumor necrosis factor α (TNFα) [26]. In contrast, IL-4 and IL-10 limit the inflammatory hyperalgesia by inhibiting the production of the former cytokines and of prostaglandins [27]. In this context, we have demonstrated that the analgesic action of glucocorticoids and of lipocortin-1 results from inhibition of the formation of hyperalgesic cytokines and prostaglandins (discussed later).

The double role of bradykinin in inflammatory pain

Bradykinin is a commonly employed mediator of pain in laboratory experiments because of its ability to produce short overt pain in humans when instilled into the base of cantharidin blisters [28], or when injected into the abdominal cavity [29] or into cephalic and brachial veins previously sensitised with serotonin [30]. These observations in humans have been supported by behavioural and electrophysiological studies [31, 32]. Work from various laboratories, including our own, has shown that in addition to its ability to directly induce nociception, bradykinin may also contribute indirectly to inflammatory hyperalgesia by stimulating the release of prostaglandins and sympathomimetic amines through the action of hyperalgesic cytokines. Thus, in the hyperalgesia induced by carrageenan or *Escherichia coli* lipolysaccharide (LPS), the release of TNFα is preceded by bradykinin generation, although at a high concentration of LPS, the importance of bradykinin is overshadowed by the direct release of TNFα by macrophages [33]. In some cases, such as following extensive tissue damage, prostaglandins may be released by injured cells without the involvement of cytokines [33].

Cyclooxygenases and inflammatory pain

The discovery in the early 1970s that aspirin and related drugs inhibit prostaglandin biosynthesis provided insight into the mechanism of action of these drugs. For many years, it was thought that prostaglandin synthase, now known as cyclooxygenase (COX), was constitutively expressed in all tissues and that the variable sensitivity of COX to non-steroidal anti-inflammatory drugs (NSAIDs) in different organs could be attributed to isoforms of the enzyme [34]. COX activity was subsequently found to be increased in tissues affected by carrageenan-induced inflammation and the activity of the enzyme could be induced by inflammatory cytokines *in vitro* [35–38]. These data suggested the existence of an inducible form of the enzyme. The two isoforms of COX (COX-1 and COX-2, representing the constitutive and inducible forms, respectively) have since been purified and cloned [39–41]. COX-1 is constitutively expressed in most tissues, including stomach, kidney, vascular smooth muscle and platelets, whereas proinflammatory COX-2 is normally undetectable in most tissues but can be induced by inflammatory cytokines such as IL-1β and TNFα in resident and migrating cells [42, 43]. The variations in the sensitivity of the two isoforms to inhibition by NSAIDs partly reflects their differing affinities for these drugs [44, 45]. The therapeutic benefits of NSAIDs derive from the inhibition of COX-2, whereas the toxic side effects are associated with the inhibition of COX-1 [45, 46]. Thus, drugs such as SC-58125, NS-398, L-745,337 and meloxicam, which have a relatively greater selectivity for COX-2, inhibit oedema and nociception *in vivo*, without causing gastric mucosal erosions [42, 47–50].

We have studied the involvement of COX-2 in inflammatory hyperalgesia using a rat paw pressure test which is a modification of the classic Randall-Sellito test [5]. When injected into naive paws, arachidonic acid produces no hyperalgesia but potentiates the hyperalgesic response to carrageenan and IL-1, both of which are known to induce COX-2. In constrast, arachidonic acid does not potentiate the hyperalgesia induced by either IL-8, a cytokine which specifically stimulates the release of sympathomimetic amines, or by the direct administration of PGE$_2$, a manoeuver which bypasses prostaglandin biosynthesis (Fig. 1). These results indicate that the amount of COX-1 present in naive paws is insufficient to produce enough prostaglandins to cause hyperalgesia but that arachidonic acid may be a useful substrate for potentiating hyperalgesia when there is the participation of COX-2. IL-1β has been reported to induce COX-2 expression [43, 51] and COX-2 has been detected in inflamed paws injected with carrageenan [42] as well as in lumbar spinal cord following the intraplantar administration of Freund's complete adjuvant [52]. The relative contribution of COX-2 to inflammatory hyperalgesia may however depend on the model investigated [53].

The induction of rat paw hyperalgesia by IL-1 can be blocked by pretreating the animals with dexamethasone, a glucocorticoid, or with lipocortin-1$_{2-26}$. This inhibition is abolished by pretreating the animals with anti-lipocortin-1 antibodies. Dexamethasone and lipocortin-1$_{2-26}$ also inhibit the release of PGE$_2$ by macrophages *in vitro* [54]. These observations suggest that the analgesic action of glucocorticoids is mediated by lipocortin-1, and that it results, at least partially, from the inhibition of prostaglandin release, possibly through the blockade of COX-2 induc-

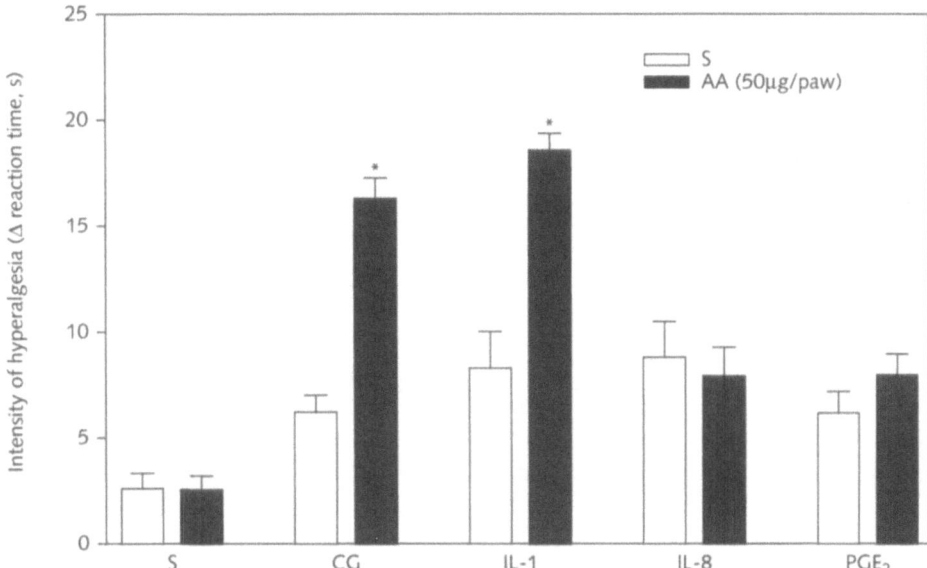

Figure 1
*The influence of arachidonic acid on the hyperalgesic response to carrageenan (CG), inter-
leukin-1β (IL-1), interleukin-8 (IL-8) and prostaglandin E_2 (PGE$_2$).*
*The columns represent the intensity of hyperalgesia in rat paws after the intraplantar injec-
tion of saline (S, 100 μl), carrageenan (CG, 50 μg/100 μl), IL-1β (IL-1, 0.25 pg/100 μl), IL-
8 (0.05 pg/100 μl) and PGE$_2$ (50 ng/100 μl) alone (empty columns) or in association with
arachidonic acid (AA, 50 μg/50 μl) (filled columns). The intensity of hyperalgesia was measured 3 h after injection of the stimuli, using a modified Randall-Sellito test in which a constant pressure of 20 mmHg was applied to the paw and discontinued (reaction time) when
the animals exhibited a freezing reaction [5]. The intensity of hyperalgesia was calculated as
the difference in the reaction times (Δ reaction time in seconds, s) obtained by subtracting
the reaction time measured 3 h after administration of the hyperalgesic substance(s) from
the control reaction time obtained prior to the injection of these same substance(s). Each column represents the mean ± standard error of mean (s.e.m.) of five rats. *p < 0.01 relative to
rats injected with saline (S).*

tion. In agreement with this, dexamethasone is reported to inhibit the induction of
COX-2 [51]. Similarly, potentiation of the hyperalgesic effect of IL-1 by arachidonic acid is abolished by dexamethasone as well as by lipocortin-1$_{2-26}$ (Fig. 2).

Meloxicam is considered to be largely selective for COX-2 [48]. Figure 3 shows
that this drug dose-dependently inhibits carrageenan- and IL-1-induced hyperalgesia without affecting that produced by PGE$_2$. Meloxicam also inhibits the potentia-

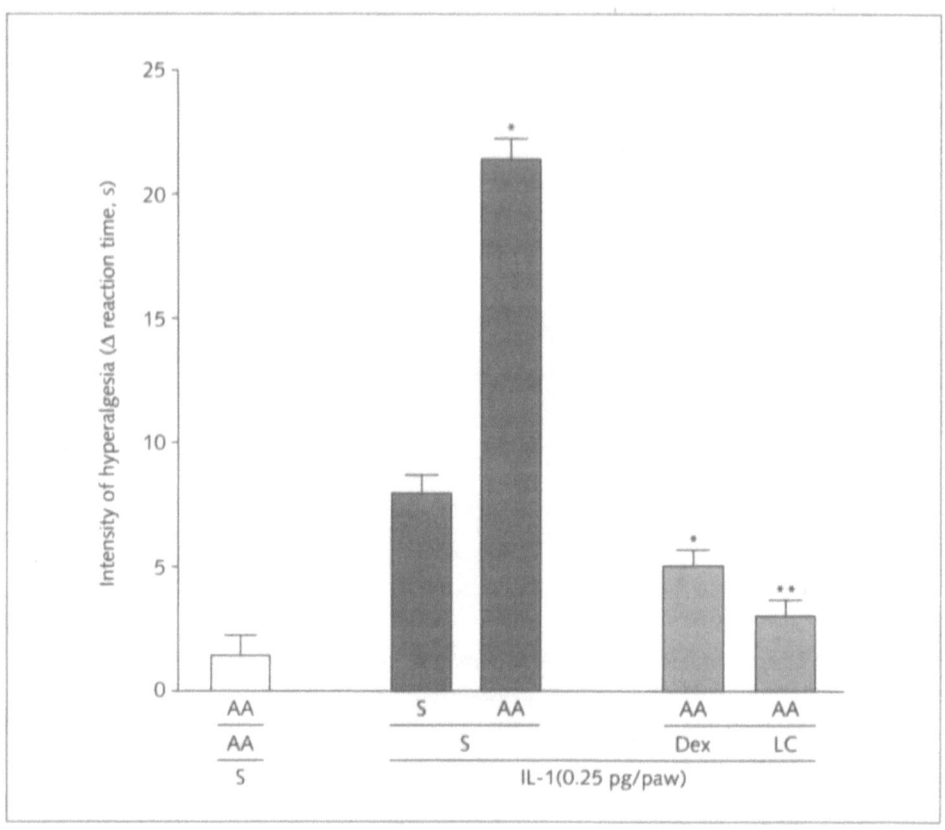

Figure 2

Inhibition by dexamethasone and lipocortin-1$_{2-26}$ of arachidonic acid-induced potentiation of the hyperalgesic effect of IL-1β.

Rats were injected with saline (S, 100 μl, i.pl.), lipocortin-1$_{2-26}$ (LC, 100 μg/50 μl, i.pl.) or dexamethasone (Dex, 5 μg/100 μl, i.pl.). Thirty minutes later, arachidonic acid (AA, 50 μg/50 μl, i.pl.) or saline (S, 50 μl, i.pl.) was injected and, after a further 10 min, IL-1β (IL-1, 0.25 pg/100 μl, i.pl.) or saline (S, 100 μl, i.pl.) was administered. The hyperalgesia was determined 3 h after injection of the hyperalgesic stimuli, as described for Figure 1. Each column represents the mean ± s.e.m. of five rats. *p < 0.001 relative to rats injected with S + IL-1. **p < 0.001 relative to rats injected with AA + IL-1. (Data from [54].)

tion by arachidonic acid of the hyperalgesia induced by both carrageenan and IL-1β. These results agree with the hypothesis that during inflammation prostaglandins are generated by an inducible COX. They also strengthen the arguments for the search for new and more specific COX-2 inhibitors [55–59] that may be useful in the treatment of pain.

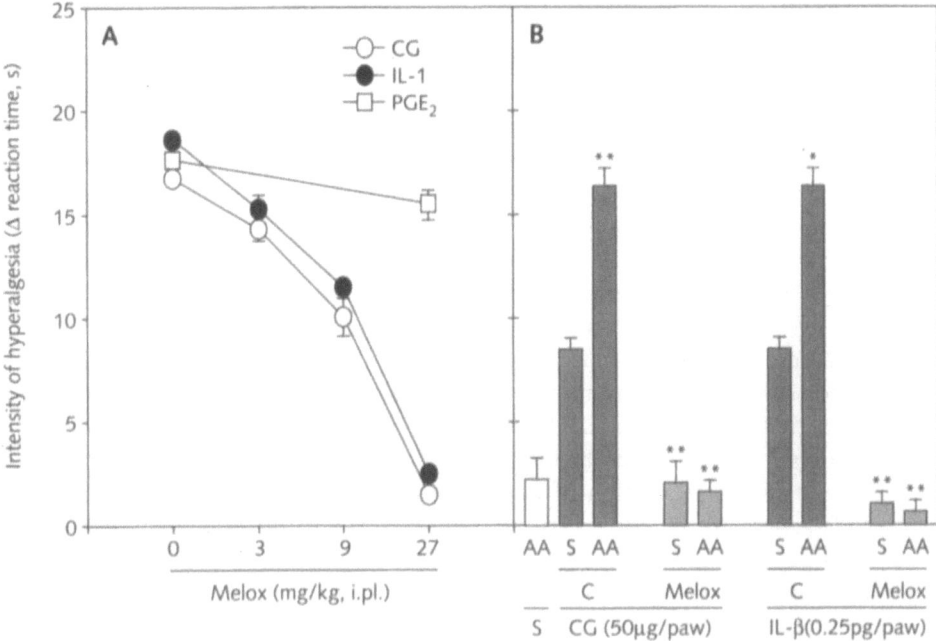

Figure 3
Panel A: Dose-dependent effect of meloxicam on the hyperalgesia induced by carrageenan,
II-1β and prostaglandin E₂. Rats were treated with the indicated doses of meloxicam 1 h
before the injection of carrageenan (CG, 100 µg/100 µl), IL-1β (0.5 pg/100 µl) or
prostaglandin E₂ (PGE₂, 100 ng/100 µl). The hyperalgesia was determined 3 h after admin-
istration of the hyperalgesic substances as described for Figure 1.
Panel B: Inhibition by meloxicam of arachidonic acid-induced potentiation of the hyperal-
gesic response to carrageenan and IL-1β. Rats were injected with saline (C, control, 100 µl,
i.p.) or meloxicam (Melox, 27 mg/kg, i.p.). Thirty minutes later, arachidonic acid (AA,
50 µg/50 µl, i.pl.) or saline (S, 50 µl, i.pl.) was injected and, after a further 10 min, car-
rageenan (50 µg/100 µl, i.pl.), IL-1β (IL-1, 0.25 pg/100 µl, i.pl.) or saline (S, 100 µl, i.pl.)
was administered. The hyperalgesia was determined 3 h after injection of the hyperalgesic
stimuli, as described for Figure 1. In both panels, the columns represent the mean ± s.e.m.
of five rats. *p < 0.001 relative to rats injected with saline (S, i.pl.). **p < 0.001 relative to
rats injected with saline (C, i.p.).

NO synthase and inflammatory pain

Nitric oxide (NO), derived from molecular oxygen and the guanidino nitrogen of L-arginine, is involved in a variety of physiological and pathological processes [60]. The formation of NO is catalysed by the enzyme NO synthase (NOS), of which constitutive (cNOS) and inducible (iNOS) forms have been identified [61]. The latter form is detected in cells only following exposure to proinflammatory agents such as cytokines and endotoxins. The activity of both enzymes can be specifically and stoichiometrically inhibited by structural analogues of L-arginine such as L-N^G-monomethyl arginine (L-NMMA) [61]. Constitutive NOS is a Ca^{2+}/calmodulin- and NADPH-dependent enzyme that is present in vascular endothelium, platelets and brain. Constitutive NOS is involved primarily in modulating physiological processes such as endothelium-dependent vascular relaxation, platelet aggregation and certain forms of central and peripheral neurotransmission. Inducible NOS can be induced by cytokines and LPS in a large variety of cells, including macrophages, hepatocytes, vascular smooth muscle and endothelial cells, and glial cells [60, 61]. The induction of iNOS in neurons has also been described [62, 63]. In contrast to cNOS, the activity of iNOS is Ca^{2+}-independent and its induction is inhibited by glucocorticoids [64]. The microbicidal and tumoricidal activities of macrophages are dependent on the generation of large amounts of NO by iNOS in these cells. Nitric oxide generated by iNOS also contributes to the cardiovascular failure seen during endotoxemic shock, as well as to the tissue lesions in inflammatory diseases and to neuronal damage [62, 65].

Role of NO in hyperalgesia

Analgesics which are cyclooxygenase inhibitors do not block ongoing hyperalgesia or the nociceptor sensitisation caused by hyperalgesic mediators such as PGE$_2$, PGI$_2$, or dopamine. However, ongoing hyperalgesia can be counteracted by the local administration of dibutyryl cGMP, or of substances able to increase intracellular cGMP levels, including acetylcholine [6–8, 11, 13], opiates [16], or analgesics such as dipyrone and diclofenac [66, 67]. The advent of drugs which either interfere with NO synthesis (NOS inhibitors) or reproduce the effects of NO (NO donors) has allowed investigation of the importance of NO in the action of directly acting analgesics.

Based on the observation that NO donors cause analgesia and that this response can be attenuated by inhibitors of soluble guanylate cyclase, it has been concluded that NO exerts its antihyperalgesic action by stimulating the formation of cGMP. Figure 4 shows that sodium nitroprusside (SNP) or 3-morpholino-sydnonimine (SIN-1) block the hyperalgesia induced by PGE$_2$. The antihyperalgesic effect of these NO donors is inhibited by methylene blue (MB), a soluble guanylate cyclase inhibitor, but not by the NOS inhibitor L-NMMA. Since the administration of NO donors causes immediate overt pain in humans [68], the effect of the NO donors in Figure 4 was measured 30 min after their administration in order to avoid an immediate effect unrelated to NO release.

Acetylcholine is known to stimulate cGMP formation in several biological systems [69]. Figure 5 shows that the analgesic effect of acetylcholine is inhibited by L-

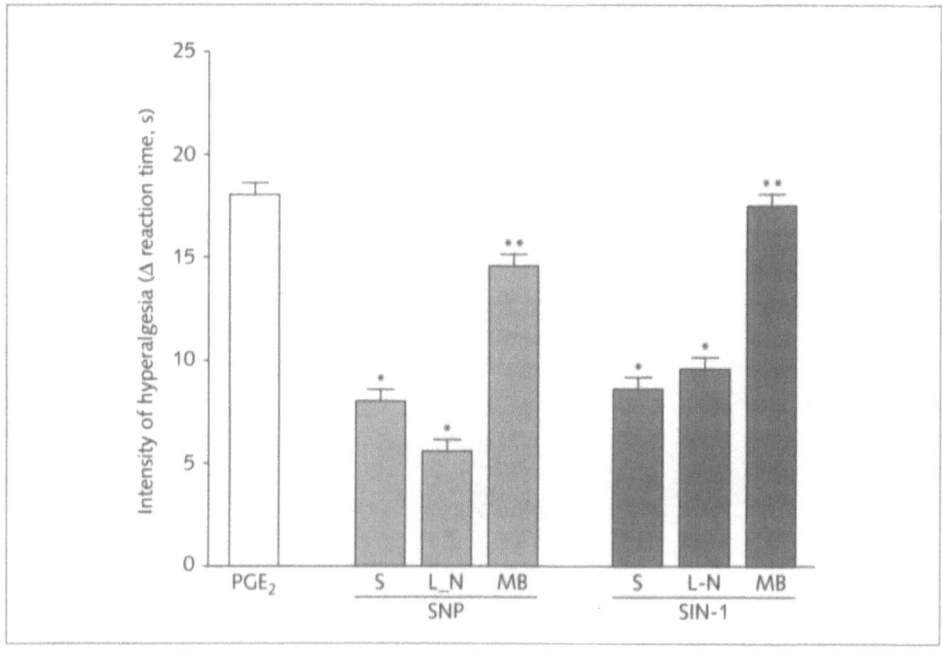

Figure 4
The influence of L-NMMA and methylene blue on the analgesic effect of the NO donors sodium nitroprusside (SNP) and 3-morpholino-sydnonimine (SIN-1).
Rats were injected with prostaglandin E_2 (PGE$_2$, 100 ng/100 µl, i.pl.), and 1 h later received saline (S, 100 µl, i.pl.), L-NMMA (L-N, 50 µg/100 µl, i.pl.) or methylene blue (MB, 500 µg/100 µl, i.pl.). After a further hour, SNP (500 µg/100 µl, i.pl.) or SIN-1 (200 µg/ 100 µl, i.pl.) was injected. The hyperalgesia was determined 3 h after the injection of PGE$_2$ as described for Figure 1. Each column represents the mean ± s.e.m. of five rats. *p < 0.001 relative to rats injected with PGE$_2$. **p < 0.001 relative to rats injected with saline (S). (Data from [13, 83].)

NMMA and by MB, but not by D-NMMA, the inactive enantiomer of L-NMMA. This figure also shows that the small analgesic effect of a low dose of acetylcholine can be enhanced by MY5445, an inhibitor of the enzyme cGMP phosphodiesterase that converts cGMP to GMP. These results strongly suggest that the analgesic action of acetylcholine is mediated by NO which increases the intracellular GMP concentration via the activation of soluble guanylate cyclase.

Several observations indicate that the L-arginine/NO/cGMP pathway has a peripheral hyperalgesic rather than analgesic effect. Thus, the intraplantar or systemic administration of N$^\omega$-nitro-L-arginine methyl ester (L-NAME, another NOS inhibitor), but not D-NAME, has been reported to produce dose-dependent

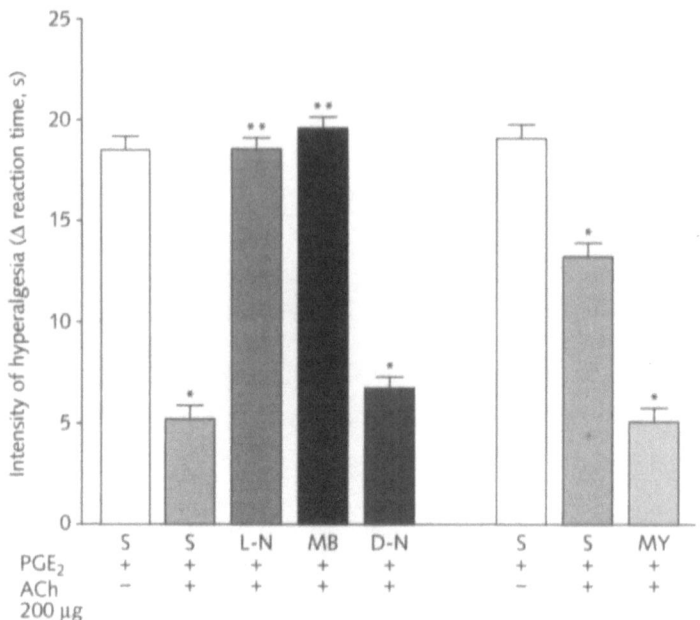

Figure 5
The influence of L-NMMA, D-NMMA, methylene blue and MY5445 on the antinociceptive action of acetylcholine.
Rats were injected with prostaglandin E_2 (PGE$_2$, 100 ng/100 µl, i.pl.) and 1 h later received saline (S, 100 µl, i.pl.), L-NMMA (L-N, 4.5 mg/kg, i.p.), methylene blue (MB, 500 µg/100 ml, i.pl.), D-NMMA (4.5 mg/kg, i.p.) or MY-5445 (MY, 50 µg/100 µl, i.pl.). After a further hour, acetylcholine (ACh, 50 µg or 200 µg/100 µl, i.pl.) was injected. The hyperalgesia was determined 3 h after the injection of PGE2 as described for Figure 1. Each column represents the mean ± s.e.m. of five rats. *p < 0.001 relative to rats injected with PGE$_2$. **p < 0.001 relative to rats injected with saline (S). (data from [13]).

antinociception in the second phase of the formalin test in rats [70]. A nociceptive role for the L-arginine/NO/cGMP pathway has also been demonstrated using other tests, such as the tail-flick and hot-plate tests, acetic acid- or phenyl-ρ-quinone-induced writhing and formalin-induced paw licking in mice [70–76]. The simplest explanation for these conflicting observations may be that the role and importance of this pathway varies among the groups of primary sensory neurones mobilised by different types of nociceptive stimuli. Alternatively, the use of different NO synthase inhibitors may yield divergent results [77]. L-NAME, in particular, is a rather pecu-

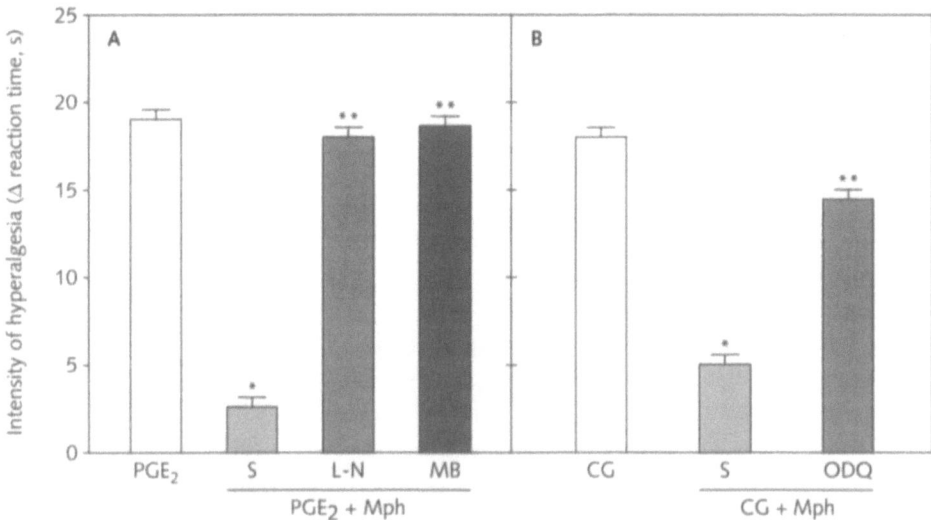

Figure 6
Effect of L-NMMA, methylene blue and ODQ on the analgesic action of morphine.
Rats were injected with prostaglandin E₂ (PGE₂, 100 ng/100 μl, i.pl., Panel A) or carrageenan
(CG, 100 μg/100 μl, i.pl., Panel B) and 1 h later received saline (S, 100 μl, i.pl.), L-NMMA
(L-N, 50 μg/100 μl, i.pl.), methylene blue (MB, 500 μg/100 μl, i.pl.) or ODQ (4 μg/100 μl,
i.pl.). After a further hour, morphine (8 μg/100 μl, i.pl.) was injected. The hyperalgesia was
determined 3 h after the injection of PGE₂ or Cg as described for Figure 1. Each column rep-
*resents the mean ± s.e.m. of five rats. *p < 0.001 relative to rats injected with PGE₂ or Cg.*
***p < 0.001 relative to rats injected with saline (S).*

liar compound since it has no inhibitory effect on some NOS but induces iNOS and stimulates soluble guanylate cyclase in neural tissue [78]. In our mechanical test of hyperalgesia induced by PGE₂, L-NAME produces analgesia which can be antago-nised by pretreating the paws with L-NMMA (IDG Duarte and SW Ferreira, unpub-lished observations).

We have proposed that peripheral opiates [16], as well as some recognised anal-gesics such as dipyrone or diclofenac [66, 67, 79], cause analgesia via stimulation of the L-arginine/NO/cGMP pathway. This proposal was based on the observation that their analgesic action was inhibited by L-NMMA and MB, and potentiated by a cGMP phosphodiesterase inhibitor.   illustrates that the analgesic effect of locally administered morphine on hyperalgesia induced by PGE₂ is inhibited by the coadministration of L-NMMA or MB (Panel A). In addition, the analgesic action of morphine on the hyperalgesia induced by carrageenan is inhibited by ODQ [80], a specific inhibitor of soluble guanylate cyclase (Panel B). The involvement of NO in

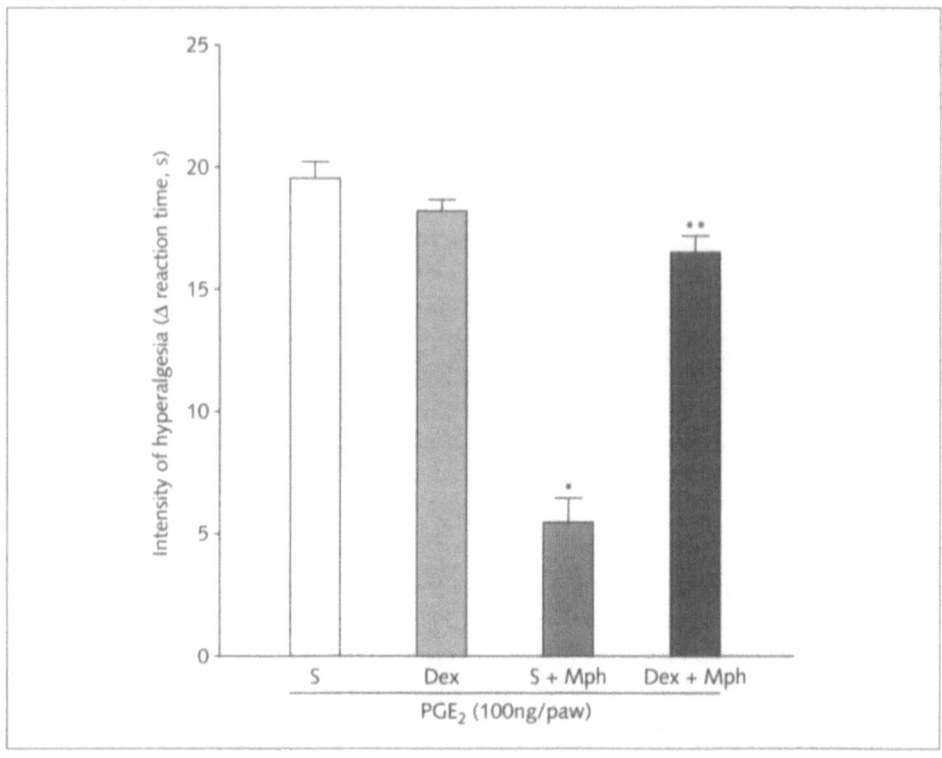

Figure 7
Inhibition by dexamethasone of the peripheral analgesic action of morphine.
Rats were treated with saline (S, 100 µl, i.pl.) or dexamethasone (Dex, 4.5 mg/kg, s.c.). One hour later, prostaglandin E₂ (PGE₂, 100 ng/100 µl, i.pl.) was injected and, after a further 2 h, morphine (Mph, 6 µg/100 µl, i.pl.) was administered to saline (S + Mph) or dexamethasone (Dex + Mph) pretreated rats. The hyperalgesia was determined 3 h after the injection of PGE₂ as described for Figure 1. Each column represents the mean ± s.e.m. of five rats. *p < 0.001 relative to rats injected with PGE₂. **p < 0.001 relative to rats pretreated with Mph i.pl.

the analgesic effect of morphine is also seen in the formalin test. Thus, morphine administered 20 min before the formalin reduces the number of flinches in the second phase of the test by 50–60% when the concentration of the irritant is 1% (v/v), but has no effect when the concentration of formalin is 5%. This finding indicates that the results obtained with the formalin test are highly dependent on the concentration of formalin and of the agent tested [73]. Pretreating the paws with either L-NMMA or MB 30 min before giving formalin does not alter the formalin (1%) response but abolishes the antinociceptive effect of morphine [81]. Thus, it is plausible that in a manner similar to acetylcholine, the molecular mechanism of the peripheral analgesia induced by morphine, dipyrone and diclofenac is dependent on the induction of the L-arginine/NO/cGMP pathway [16, 66, 67, 79]. The finding that morphine stimulates cGMP formation [82] is in line with our hypothesis.

The NOS which is stimulated by peripherally acting analgesics seems to be the inducible isoform since although L-arginine has no direct antihyperalgesic effect on PGE2-induced hyperalgesia, it potentiates the analgesic action of morphine. Moreover, the local analgesic action of morphine on PGE_2-induced hyperalgesia is blocked by dexamethasone (Fig. 7) which, although it does not affect the hyperalgesia induced by PGE_2, is known to inhibit the induction of NOS [54, 65]. The analgesic response to dipyrone or morphine given intrathecally is also blocked by pretreating the animals with dexamethasone (unpublished observations). To date, little is known about the cellular origin of the NO which affects primary sensory neurones. However, the participation of neuronal NOS is suggested by the finding that the analgesia obtained following the intrathecal injection of morphine or dipyrone can be blocked by the intraplantar injection of L-NMMA or MB (unpublished observations). These findings indicate that intrathecally injected analgesics act on primary sensory neurones which may themselves be the main source of this NO. Indeed, the induction of iNOS in these neurones has been reported [62, 63].

In conclusion, the findings discussed above indicate that the prostaglandin which mediates hyperalgesia, and the NO which stimulates the formation of neuronal cGMP to promote analgesia appear to be products of the inducible isoforms of COX and NOS, respectively.

Acknowledgements
The authors thank Ms I.R. Santos for technical assistance. This work was supported by grants from FAPESP and CNPq (Brazil).

References

1   Handwerker HO (1976) Influences of algogenic substances on the discharges of unmyelinated cutaneous nerve fibres identified as nociceptor. In: JJ Bonica, D Albe-Fessard (eds): *Advances in pain research and therapy*. Raven Press, New York, Vol. 1, 41–45

2   Perl ER (1976) Sensitization of nociceptors and its relation to sensation. In: JJ Bonica, D Albe-Fessard (eds): *Advances in pain research and therapy*. Raven Press, New York, Vol. 1, 17–34

3   McMahon SB, Koltzenburg M (1990) Novel classes of nociceptors: beyond Sherrington. TINS 13: 199–201

4   Messlinger K (1997) What is a nociceptor? *Anaesthesist* 46: 142–153

5   Ferreira SH, Lorenzetti BB, Correa FMA (1978) Central and peripheral antialgesic action of aspirin-like drugs. *Eur J Pharmacol* 53: 39–48

6   Ferreira SH, Nakamura M (1979) I – Prostaglandin hyperalgesia: a cAMP/Ca++ dependent process. *Prostaglandins* 18: 179–190

7   Taiwo YO, Bjerknes LK, Goetz EJ, Levine JD (1989) Mediation of primary afferent peripheral hyperalgesia by the cAMP second messenger system. *Neuroscience* 32: 577–580

8   Follenfant RL, Nakamura M, Garland LG (1990) Sustained hyperalgesia in rats evoked by the protein kinase inhibitor H-7. *Br J Pharmacol* 99: 289P

9   Kress M, Rodl J, Reech PW (1996) Stable analogues of cyclic AMP but not cyclic GMP sensitize unmyelinated primary afferents in rat skin to heat stimulation but not to inflammatory mediators, *in vitro*. *Neuroscience* 74: 609–617

10   Wang JF, Khasar SG, Ahlgren SC, Levine JD (1996) Sensitization of C-fibers by prostaglandin E2 in the rat is inhibited by guanosine 5'-O-(2-thiodiphosphate), 2',5'-dideoxyadenosine and Walsh inhibitor peptide. *Neuroscience* 71: 259–263

11   Ferreira SH, Lorenzetti BB (1994) Glutamate spinal retrograde sensitization of primary sensory neurons associated with nociception. *Neuropharmacology* 33: 1479–1485

12   Gold MS, Reichling DB, Shuster MJ, Levine JD (1996) Hyperalgesic agents increase a tetrodotoxin-resistant Na+ current in nociceptors. *Proc Natl Acad Sci USA* 93: 1108–1112

13   Duarte IDG, Lorenzetti BB, Ferreira SH (1990) Peripheral analgesia and activation of the nitric oxide-cyclic GMP pathway. *Eur J Pharmacol* 186: 289–293

14   Murad F (1986) Cyclic guanosine monophosphate as a mediator of vasodilation. J Clin Invest 78: 1–5

15   Palmer RMJ, Ferrige AG, Moncada S (1987) Nitric oxide release accounts for the biological activity of endothelium-derived relaxing factor. *Nature* 327: 524–526

16   Ferreira SH, Duarte IDG, Lorenzetti BB (1991) Molecular basis of acetylcholine and morphine analgesia. In: MJ Parnham, MA Bray, WB van den Berg (eds): *Drugs in inflammation*. Birkhäuser Verlag, Basel, 101–106

17   Ferreira SH (1983) Prostaglandins, peripheral and central analgesia. In: JJ Bonica (ed): *Advances in pain research and therapy*. Raven Press, New York, Vol. 5, 627–634

18   Nakamura M, Ferreira SH (1987) A peripheral sympathetic component in inflammatory hyperalgesia. *Eur J Pharmacol* 135: 145–153

19   Coderre TJ, Abbott FV, Melzack R (1984) Effects of peripheral antisympathetic treatments in the tail-flick, formalin and autotomy tests. Pain 18: 13–23

20   Duarte IDG, Nakamura M, Ferreira SH (1988) Participation of the sympathetic system in acetic acid-induced writhing in mice. *Brazilian J Med Biol Res* 21:341–343

21   Tracey DJ, Cunningham JE, Romm MA (1995) Peripheral hyperalgesia in experimental neuropathy: mediation by alpha 2-adrenoreceptors on post-glanglionic sympathetic terminals. *Pain* 60: 317–327

22   Raja SN (1995) Role of the sympathetic nervous system in acute pain and inflammation. *Ann Med* 27: 241–246

23   Ferreira SH (1980) Are macrophages the body's alarm cells? *Agents Actions* 10: 229–230

24   Ferreira SH, Lorenzetti BB, Bristow AF, Poole S (1988) Interleukin-1β as a potent hyperalgesic agent antagonized by a tripeptide analogue. *Nature* 334: 698–700

25   Cunha FQ, Lorenzetti BB, Poole S, Ferreira SH (1991) Interleukin 8 as a mediator of sympathetic pain. *Br J Pharmacol* 104: 765–767

26   Cunha FQ, Poole S, Lorenzetti BB, Ferreira SH (1992) The pivotal role of tumor necrosis factor α in the development of inflammatory hyperalgesia. *Br J Pharmacol* 107: 660–664

27   Poole S, Cunha FQ, Selkirk S, Lorenzetti BB, Ferreira SH (1995) Cytokine-mediated inflammatory hyperalgesia limited by interleukin-10. *Br J Pharmacol* 115: 684–688

28   Whalley ET, Clegg S, Stewart JM, Vavrek RJ (1987). The effect of kinin agonist and antagonists on the pain response of the human blister base. *Naunyn-Schemied Arch Pharmacol* 336: 652–655

29   Lim RKS, Miller DG, Guzman F (1967) Pain and analgesia evaluated by intraperitoneal bradykinin-evoked pain method in man. *Clin Pharmacol Ther* 8: 521–542

30   Sicuteri F, Franciullacci FM, Franchi G, Del Bianco PL (1965) Serotonin-bradykinin potentiation of the pain receptors in man. *Life Sci* 4: 309–316

31   Dray A, Bettaney J, Forster P, Perkin MN (1988). Activation of a bradykinin receptor in peripheral nerve and spinal cord in the neonatal rat *in vitro*. *Br J Pharmacol* 95: 1008–1010

32   Lang E, Novak A, Reeh PW, Handwerker HO (1990). Chemosensitivity of fine afferents from rat skin *in vitro*. *J Neurophysiol* 63: 887–901

33   Ferreira SH, Lorenzetti BB, Poole S (1993) Bradykinin initiates cytokine-mediated inflammatory hyperalgesia. *Br J Pharmacol* 110: 1227–1231

34   Flower RJ, Vane JR (1972) Inhibition of prostaglandin synthetase in brain explains the anti-pyretic activity of paracetamol (4-acetamido-phenol). *Nature (Lond.)* 240: 410–411

35   Raz A, Wyche A, Siegel N, Needleman P (1988) Regulation of fibroblast cyclooxygenase synthesis by interleukin-1. *J Biol Chem* 263: 3022–3028

36   Fu J, Masferrer JL, Seibert K, Raz A, Needleman P (1990) The induction and suppression of prostaglandin H2 synthase (cyclooxygenase) in human monocytes. *J Biol Chem* 265: 737–740

37   Masferrer JL, Zweifel BS, Seibert K, Needleman P (1992) Selective regulation of cellular cyclooxygenase by dexamethasone and endotoxin in mice. *J Clin Invest* 86: 1375–1379

38   Sano H, Hla TM, Maier JA, Crofford LJ, Case JP, Maciag T, Wilder RL (1992) *In vivo* cyclooxygenase expression in synovial tissues of patients with rheumatoid arthritis and osteoarthritis and rats with adjuvant and streptococcal cell wall arthritis. *J Invest Clin* 89: 97–108

39 Langenbach R, Morham SG, Tiano HF, Loftin CD, Ghanayem BI, Chulada PC, Maher JF, Lee CA, Goulding EH, Kluckman KD et al (1995) Prostaglandin synthase 1 gene disruption in mice reduces arachidonic acid-induced inflammation and indomethacin-induced gastric ulceration. *Cell* 83: 483–492

40 Morham SG, Langenbach R, Loftin CD, Tiano HF, Vouloumanos N, Jennette JC, Maher JF, Kluckman KD, Ledford A, Lee CA et al (1995) Prostaglandin synthase 2 gene disruption causes severe renal pathology in the mouse. *Cell* 83: 473–482

41 Dinchuk JE, Car BD, Focht RJ, Johnston JJ, Jaffee BD, Covington MB, Contel NR, Eng VM, Collins RJ, Czerniak PM et al (1995) Renal abnormalities and an altered inflammatory response in mice lacking cyclooxygenase II. *Nature* 378: 406–409

42 Seibert K, Zhang Y, Leahy K, Hauser S, Masferrer J, Perkins W, Lee L, Isakson P (1994) Pharmacological and biochemical demonstration of the role of cyclooxygenase 2 in inflammation and pain. *Proc Natl Acad Sci USA* 91: 12013–12017

43 Bakhle YS, Botting RM (1996) Cyclooxygenase-2 and its regulation in inflammation. *Mediat Inflammat* 5: 305–323

44 Kargman S, Wong E, Greig GM, Falgueyret JP, Cromlish W, Ethier D, Yergey JA, Riendeau D, Evans JF, Kennedy B et al (1996) Mechanism of selective inhibition of human prostaglandin G/H synthase-1 and synthase-2 in intact cells. *Biochem Pharmacol* 52: 1113–1125

45 Smith WL, DeWitt DL (1995) Biochemistry of prostaglandin endoperoxide H synthase-1 and synthase-2 and their differential susceptibility to nonsteroidal anti-inflammatory drugs. *Semin Nephrol* 15: 179–194

46 Seibert K, Masferrer J, Zhang Y, Gregory S, Olson G, Hauser S, Leahy K, Perkins W, Isakson P (1995) Mediation of inflammation by cyclooxygenase-2. *Agents Actions* (Suppl.) 46: 41–50

47 Chan CC, Boyce S, Brideau C, Ford-Hutchinson AW, Gordon R, Guay D, Hill RG, Li CS (1995) Pharmacology of a selective cyclooxygenase-2 inhibitor, L-745,337: a novel nonsteriodal anti-inflammatory agent with an ulcerogenic sparing effect in rat and non-human primate stomach. *J Pharmacol Exp Ther* 274: 1531–1537

48 Engelhardt G (1996) Pharmacology of meloxicam, a new non-steriodal anti-inflammatory drug with an improved safety profile through preferential inhibition of COX-2. *Br J Rheumatol* 35: 4–12

49 Futaki N, Yoshikawa K, Hamasaka Y, Arai I, Higuchi S, Iizuka H, Otomo S (1993) NS-398, a novel non-steroidal anti-inflammatory drug with potent analgesic and antipyretic effects, which causes minimal stomach lesions. *Gen Pharmacol* 24: 105–110

50 Futaki N, Takahashi S, Yokoyama M, Arai I Higuchi S, Otomo S (1994) NS-398, a new anti-inflammatory agent, selectively inhibits prostaglandin G/H synthase/cyclooxygenase (COX-2) activity *in vitro*. *Prostaglandins* 47: 55–59

51 Mitchell JA, Belvisi MG, Akarasereenont P, Robbins RA, Kwon OJ, Croxtall JJ, Barnes PJ, Vane JR (1994) Induction of cyclo-oxygenase-2 by cytokines in human pulmonary epithelial cells: regulation by dexamethasone. *Br J Pharmacol* 113: 1008–1014

52 Hay CH, Trevethick MA, Wheeldon A, Bowers JS, de Belleroche JS (1997) The poten-

tial role of spinal cord cyclooxygenase-2 in the development of Freund's complete adjuvant-induced changes in hyperalgesia and allodynia. *Neuroscience* 78: 843–850

53   Yamamoto T, Nozaki-Taguchi N (1996) Analysis of the effects of cyclooxygenase (COX)-1 and COX-2 in spinal nociceptive transmission using indomethacin, a nonselective COX inhibitor, and NS-398, a COX-2 selective inhibitor. *Brain Res* 739: 104–110

54   Ferreira SH, Cunha FQ, Lorenzetti BB, Michelin MA, Perretti M, Flower RJ, Poole S (1997) Role of lipocortin-1 in the anti-hyperalgesic actions of dexamethasone. *Br J Pharmacol* 121: 883–888

55   Li CS, Black WC, Chan CC, Ford-Hutchinson AW, Gauthier JY, Gordon R, Guay D, Kargman S, Lau CK, Mancini J et al (1995) Cyclooxygenase-2 inhibitors: synthesis and pharmacological activities of 5-methanesulfonamido-1-inandone derivatives. *J Med Chem* 38: 4897–4905

56   Huang HC, Li JJ, Garland DJ, Chamberlain TS, Reinhard EJ, Manning RE, Seibert K, Koboldt CM, Gregory SA, Anderson GD et al (1996) Diarylspiro[2.4]heptenes as orally active, highly selective cyclooxygenase-2 inhibitors: synthesis and structure-activity relationships. *J Med Chem* 39: 253–266

57   Li JJ, Norton MB, Reinhard EJ, Anderson GD, Gregory SA, Isakson PC, Koboldt CM, Masferrer JL, Perkins WE, Seibert K et al (1996) Novel terphenyls as selective cyclooxygenase-2 inhibitors and orally active anti-inflammatory agents. *J Med Chem* 39: 1846–1856

58   Penning TD, Talley JJ, Bertenshaw SR, Carter JS, Collins PW, Docter S, Graneto MJ, Lee LF, Malecha JW, Miyashiro JM et al (1997) Synthesis and biological evaluation of the 1,5-diarylpyrazole class of cyclooxygenase-2 inhibitors: identification of 4-[5-(4-methylphenyl)-3-(trifluoromethyl)-1H-pyrazol-1-yl]benzenesulfonamide (SC-58635, celecoxib). *J Med Chem* 40: 1347–1365

59   Khanna IK, Weier RM, Yu Y, Xu XD, Koszyk FJ, Collins PW, Koboldt CM, Veenhuizen AW, Perkins WE, Casler JJ et al (1997) 1,2-Diarylimidazoles as potent, cyclooxygenase-2 selective, and orally active antiinflammatory agents. *J Med Chem* 40: 1634–1647

60   Moncada S, Palmer RMJ, Higgs EA (1991) Nitric oxide: physiology, pathophysiology, and pharmacology. *Pharmacol Rev* 43: 109–142

61   Fukuto JM, Chaudhuri G (1995) Inhibition of constitutive and inducible nitric oxide synthase: potential selective inhibition. *Annu Rev Pharmacol Toxicol* 35: 165–194

62   Wong ML, Rettori V, Al-Shekhlee A, Bongiorno PB, Canteros G, McCann SM, Gold PW, Licinio J (1996) Inducible nitric oxide synthase gene expression in the brain during systemic inflammation. *Nature Medicine* 2: 581–584

63   Minc-Golomb D, Tsarfaty I, Schwartz JP (1994) Expression of inducible nitric oxide synthase by neurones following exposure to endotoxin and cytokine. *Br J Pharmacol* 112: 720–722

64   Kengatharan KM, De Kimpe SJ, Thiemermann C (1996) Role of nitric oxide in the circulatory failure and organ injury in a rodent model of Gram-positive shock. *Br J Pharmacol* 119: 141–1421

65   Wright CE, Rees DD, Moncada S (1992) Protective and pathological roles of nitric oxide in endotoxin shock. *Cardiovasc Res* 26: 48–57

66   Lorenzetti BB, Ferreira SH (1996) Activation of the arginine-nitric oxide pathway in primary sensory neurons contributes to dipyrone-induced spinal and peripheral analgesia. *Inflamm Res* 45: 308–311

67   Duarte IDG, Santos IR, Lorenzetti BB, Ferreira SH (1992) Analgesia by direct antagonist of nociceptor sensitization involves the arginine-nitric oxide-cGMP pathway. *Eur J Pharmacol* 217: 225–227

68   Holthusen H, Arndt JO (1994) Nitric oxide evokes pain in humans on intracutaneous injection. *Neurosci Lett* 165: 71–74

69   Lee TP, Kuo JF, Greenhouse P (1972) Role of muscarinic cholinergic receptors in regulation of guanosine 3':5'-cyclic monophosphate content in mammalian brain muscle. *Proc Natl Acad Sci USA* 69: 3287–3291

70   Haley JE, Dickenson AH, Schachter M (1992) Electrophysiological evidence for a role of nitric oxide in prolonged chemical nociception in the rat. *Neuropharmacol*, 31: 251–258

71   Morgan CV, Babbedge RC, Gaffen Z, Wallace SL, Hart SL, Moore PK (1992) Synergistic anti-nociceptive effect of L-N^G-nitro arginine methyl ester (L-NAME) and flurbiprofen in the mouse. *Br J Pharmacol* 106: 493–497

72   Malmberg AB, Yaksh TL (1993) Spinal nitric oxide synthase inhibition blocks NMDA-induced thermal hyperalgesia and produces antinociception in the formalin test in rats. *Pain* 54: 291–300

73   Kawabata A, Manabe S, Manabe Y, Takagi H (1994) Effect of topical administration of L-arginine on formalin-induced nociception in the mouse: a dual role of peripherally formed NO in pain modulation. *Br J Pharmacol* 112: 547–550

74   Mustafa AA (1992) Mechanisms of L-N^G-nitro arginine methyl ester-induced antinociception in mice: a role for serotonergic and adrenergic neurons. *Gen Pharmacol* 23: 1177–1182

75   Moore PK, Oluyomi AO, Babbedge RC, Wallace P, Hart SL (1991) L-N^G-nitro arginine methyl ester exhibits antinociceptive activity in the mouse. Br J Pharmacol 102: 198–202

76   Meller ST, Cummings CP, Traub RJ, Gebhart GF (1994) The role of nitric oxide in the development and maintenance of the hyperalgesia produced by intraplantar injection of carrageenan in the rat. *Neuroscience* 60: 367–374

77   Babbedge RC, Hart SL, Moore PK (1993) Anti-nociceptive activity of nitric oxide synthase inhibitors in the mouse: dissociation between the effect of L-NAME and L-NMMA. *J Pharm Pharmacol* 45: 77–79

78   Miller MJS, Thompson JH, Liu X, Eloby-Childress S, Sadowska-Krowicka H, Zhang XJ, Clark DA (1996) Failure of L-NAME to cause inhibiton of nitric oxide synthesis: Role of inducible nitric oxide synthase. *Inflamm Res* 45: 272–276

79   Tonussi CR, Ferreira SH (1994) Mechanism of diclofenac analgesia: direct blockade of inflammatory sensitization. *Eur J Pharmacol* 251: 173–179

80 Moro MA, Russell RJ, Cellek S, Lizasoain I, Su Y, Darley-Usmar VM, Radomski MW, Moncada S (1996) cGMP mediates the vascular and platelet actions of nitric oxide: Confirmation using an inhibitor of the soluble guanylyl cyclase. *Proc Natl Acad Sci USA* 93: 1480–1485

81 Granados-Soto V, Rufino MO, Lopes LDG, Ferreira SH (1997) Evidence for the involvement of the L-arginine-NO-cGMP pathway in the antinociceptive effect of morphine in the formalin test. *Eur J Pharmacol; in press*

82 Minneman KP, Iversen LL (1976) Enkephalin and opiate narcotics increase cyclic GMP accumulation in slices of rat neostriatum. *Nature* 262: 313–314

83 Ferreira, SH, Duarte IDG, Lorenzetti BB (1991). The molecular mechanism of action of peripheral morphine analgesia: stimulation of the cGMP system via nitric oxide release. *Eur J Pharmacol* 201:121–122

Neuroinflammation

Bernd C. Kieseier and Hans-Peter Hartung

Department of Neurology, Karl-Franzens-Universität, Auenbruggerplatz 22, 8036 Graz, Austria

In the past, the nervous system has long been regarded as a site of immunologic privilege. This conception was based on the assumptions that (a) an anatomically tight interface between blood vessel wall and the neural parenchyma, the so called blood-brain (BBB) and blood-nerve barrier (BNB), respectively, form a strict separation between the systemic immune compartment and the nervous system, and that (b) other immunologic mechanisms such as the expression of major histocompatibility complex (MHC) molecules, T cell surveillance, and lymphatic drainage are missing. Accumulated experimental knowledge indicates that these notions are no longer tenable. We know that immune mediators occurring in the nervous system are recruited from the systemic lymphoid organs, and that endogenous cells of the nervous system, such as astrocytes, microglia or Schwann cells, play key roles in triggering and regulating the immunologic machinery in the nervous system. The model in which an aberrant immune response as a causative mechanism in neurologic diseases has been extensively studied, is known as experimental autoimmune encephalomyelitis (EAE). While it reproduces essential aspects of the pathogenesis of human multiple sclerosis (MS) [1], it was also instrumental in recognizing basic features of the organisation of the immune system [2].

EAE is an inflammatory disorder of the central nervous system (CNS). In susceptible rodents it can be actively induced by immunization with various CNS proteins, or adoptively transferred by injection of activated encephalitogenic T cells specific for these antigens [1, 3]. One of the histomorphologic hallmarks of EAE, and hence of MS, are perivascular inflammatory infiltrates, mainly consisting of T cells and macrophages, the latter being regarded as the most important effector cells [4, 5].

The animal model in which immune mediated inflammation within the peripheral nervous system (PNS) has been most thoroughly investigated, is known as experimental autoimmune neuritis (EAN). It is an acute demyelinating inflammatory polyradiculoneuropathy, sharing many similarities with the human Guillain-Barré syndrome (GBS) [6]. EAN can be induced in susceptible species by immu-

Inducible Enzymes in the Inflammatory Response, edited by D.A. Willoughby and A. Tomlinson
© 1999 Birkhäuser Verlag Basel/Switzerland

nization with whole peripheral nerve homogenate, myelin, or various myelin proteins, and by adoptive transfer of neuritogenic T cell lines [7, 8]. The disease is characterized by infiltrating lymphocytes and macrophages in the PNS, predominantly found around venules [9, 10].

Multifocal inflammation and damage to the myelin sheath are salient pathological hallmarks of both EAE/MS and EAN/GBS. However, whereas collective evidence currently points to T cells as the predominant mediators of the immune response in MS, the pathogenesis of GBS appears to be more heterogeneous [11].

Of paramount importance in the pathogenesis of inflammation within the nervous system is the transendothelial migration of leukocytes. Once activated in the periphery, circulating T cells need to cross the BBB or BNB in order to initiate a local immune response. The mechanism of transendothelial migration is a multistep process occurring in an ordered sequential fashion, encompassed by the complex interaction of a variety of adhesion molecules and cytokines [11, 12]. Once in the nervous system, CD4+ T lymphocytes undergo local reactivation and clonal expansion if they encounter the specific antigen presented by major histocompatibility complex (MHC) class II molecules and the simultaneous delivery of additional co-stimulatory signals on the cell surface of an antigen-presenting cell. These CD4+ T cells can differentiate into two types of effector T cells: (a) inflammatory T cells (T_H1) that activate macrophages/microglia to increase phagocytic activity and synthesis of inflammatory mediators by releasing the proinflammatory cytokines tumor necrosis factor α (TNFα) and interferon γ (IFNγ); (b) helper T cells (T_H2) that activate specific B cells to produce antibodies.

As previously mentioned, macrophages/microglia play a pivotal role in the inflammatory response within the nervous system (n.b. in the PNS microglial cells do not exist but tissue-resident endoneurial macrophages are thought to exhibit equivalent functions). These cells act as antigen presenters and are of critical importance in the amplification and effector phase of immune-mediated demyelination [9, 13]. Activated by cell contact and IFNγ, macrophages convert into potent effector cells of the immunologic response. Operative mechanisms of their efficacy include phagocytosis and release of proinflammatory cytokines and highly toxic mediators such as oxygen radicals, nitric oxide, and proteases. Given their antibacterial properties, release of these mediators is of paramount importance in host defense, however they are also toxic to host cells. Thus, tight regulation of macrophage activity by inflammatory T cells allows the specific deployment of this effective instrumentarium, minimizing the expense of tissue damage (Fig. 1).

One mechanism of controlling the action of such potent mediators is provided by the inducibility of enzymes involved in the regulating process of enzymatic activity. In this chapter the role of certain inducible enzymes in the pathogenesis of inflammatory diseases of the nervous system will be reviewed. Most of the data presented are based on findings from autoimmune demyelinating diseases, which does not reflect a preferential expression of these enzymes in this category of inflammatory disorders. The reason is more historical, since most of the fundamental investigations on the inflammatory response within the nervous system were carried out in this field of neurologic diseases.

Nitric oxide synthase

Function and regulation

The free radical nitric oxide (NO) is an inorganic gas that critically mediates multifarious biological functions, including vascular and muscle relaxation, platelet aggregation, neuronal-cell function, microbicidal and tumoricidal activity, and a range of immunopathologies. NO has a half life of 3–5 seconds when dissolved in biological fluids containing oxygen, and is generated from the terminal guanidino group of L-arginine and molecular oxygen by the enzyme NO synthase (NOS) [14]. At present, three principal forms of NOS proteins are known, all of which are products of different genes: the neuronal (nNOS), the macrophage or inducible (iNOS), and the endothelial (eNOS). However, these terms can be misleading since the localization of these NOS isoforms is not restricted to the name giving source: for example, iNOS does not only occur in macrophages [15] but in several cell types such as lymphocytes [16], neutrophils [17], hepatocytes [18], and fibroblasts [19], whereas eNOS is not restricted to endothelial cells of blood vessels but also exists in neurons, especially in pyramidal cells of the hippocampus, and nNOS also occurs in skeletal muscle complexed with dystrophin. Since neither eNOS nor nNOS are confined to their respective cell type and since both are constitutively expressed, these enzymes are summarized under the term cNOS (constitutive NOS). cNOS is calcium and calmodulin dependent for enzymatic activity, whereas the production of iNOS is calcium-independent, and, in contrast to the former, regulated by different cytokines and other immunological stimuli. The amount of NO synthesized in cells in which iNOS has been induced is much higher compared to cells producing NO via the constitutive enzymes.

The proinflammatory cytokines IFNγ, TNFα, and interleukin-1 (IL-1) are to date the best known inducers of iNOS transcription in macrophages and various somatic cells, including astrocytes and microglia. In contrast, transforming growth factor β (TGFβ), IL-4 and IL-10 in macrophages [20], and IL-8 in neutrophils [21], are capable of inhibiting iNOS induction. This panel of inducers and inhibitors of iNOS is mirrored in the cytokine profile secreted by T_H1 and T_H2 cells, respectively. Evidence is increasing that NO can both enhance and suppress the function of T cells, important effectors in the pathogenesis of neuroimmunologic disorders, potentially by at least two different mechanisms: NO is capable of modulating antigen presentation by downregulating the expression of MHC class II molecules on antigen-presenting cells, thus reducing T cell proliferation [22]. On the other hand, NO can

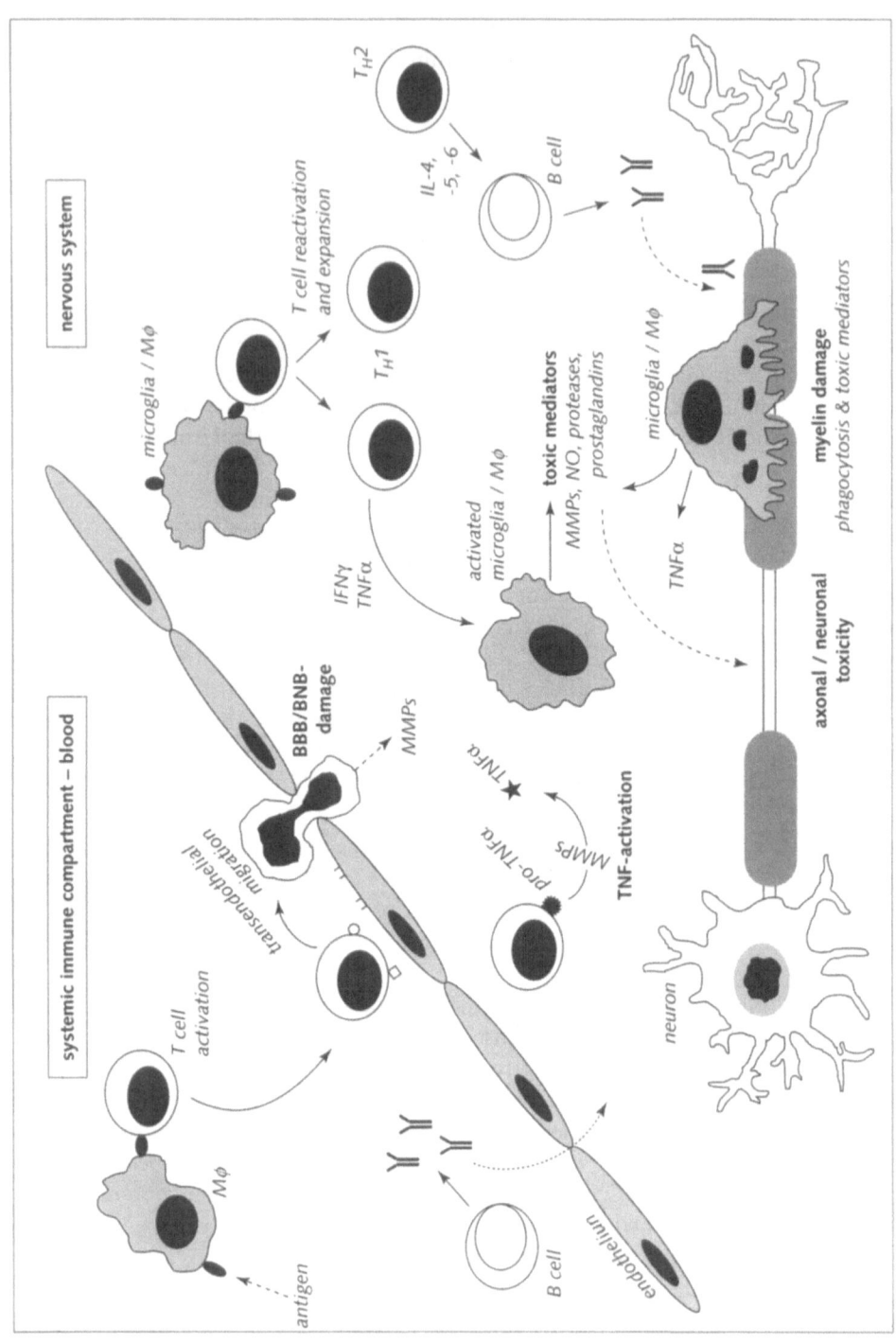

block the secretion of IL-2 and IFNγ, both autocrine mediators of T cell function, and thus inhibit the expansion of T_H1 cells, which themselves are able to produce large amounts of NO [23]. T_H2 cells, in contrast, neither produce nor does their function seem to be altered by NO. Thus, regulating T cell proliferation, NO becomes a key player in controlling the expansion of T cell subsets, mediating immunopathology. Moreover, NO can downregulate its own synthesis by inhibiting both iNOS and cNOS in a negative feedback manner via the heme moiety of NOS [24]. Finally, glucocorticoids are known to abrogate the induction, but not the activity, of iNOS [25].

The role of nitric oxide synthase in the inflamed nervous system

Until now, research has primarily focused on the constitutively expressed nNOS present in the CNS and PNS. However, the presence of iNOS in the nervous system is known. An emerging body of evidence suggests that NO plays a decisive role in the pathogenesis of inflammatory diseases of the nervous system [26]. In any inflammatory condition one would expect iNOS to be upregulated and to generate high levels of NO over extended periods of time. Indeed, several groups were able to demonstrate increased NO production via an iNOS pathway in various models of neuroinflammation [27, 28]. Furthermore, expression of iNOS has been found in demyelinating MS lesions [29] and in spinal cords of animals with EAE [30]. To test any potential role of NO in the pathogenesis of inflammatory diseases in the nervous system, the effects of NOS inhibitors on disease expression have been investigated in various rodent and murine models of EAE and EAN. The application of aminoguanidine (AG), a nucleophilic hydrazine compound, in EAE produced con-

Figure 1

The basic principles in the pathogenesis of the inflammatory immune response in the nervous system. Autoreactive T cells recognize a specific antigen, and upon activation can cross the blood-brain (BBB) and blood-nerve-barrier (BNB), respectively. Matrix metalloproteinases (MMPs) appear to be critically involved in damaging this barrier. In the nervous system the autoreactive T cells undergo local reactivation and clonal expansion if they encounter the specific antigen in context with major histocompatibility complex class II molecules and the simultaneous delivery of co-stimulatory signals on the cell surface of antigen-presenting cells. By secreting proinflammatory cytokines, T_H1 cells activate microglia cells or macrophages (Mφ) to enhance phagocytic activity, production of cytokines, and release of toxic mediators such as nitric oxide (NO), matrix metalloproteinases (MMPs), and other proteases, propagating axonal/neuronal toxicity or activation of tumor necrosis factor (TNF). Autoantibodies against myelin antigens may cross the endothelial barrier or are locally produced by B-cells, which are stimulated by T_H2 cells.

tradictory results. Initially, Cross et al. [31] reported an inhibitory effect of AG on the development of clinical signs of AT-EAE in SJL mice, data which were confirmed in AT-EAE of the Lewis rat by Zhao and co-workers [32] and recently, in myelin and T cell induced EAE in (PL/J × SJL) F_1 mice, by Brenner et al. [33]. However, Zielasek and colleagues, using the same model as Zhao et al., showed that AG treatment aggravated EAE [34]. These findings were supported by Ruuls and co-workers [35], who found EAE exacerbated by the application of N^G-monomethyl-L-arginine and nitro-L-arginine-methylester (both L-arginine analogues and NOS inhibitors). A satisfactory explanation for this discrepancy has not yet been given, and this issue certainly warrants further study. In AT-EAN however, Zielasek et al. reported that NOS inhibition resulted in a partial reduction of disease severity [34]. The above mentioned findings suggest a central role of NOS in neural inflammation. However, the exact mechanisms by which this occurs are not fully understood at present. As a free radical, NO could potentially exert direct toxic effects on many cell types within the nervous system. Merrill et al. demonstrated that the ability of microglial cells to lyse rat oligodendrocytes *in vitro* is mediated through an NO-dependent pathway [36]. On the other hand, oligodendrocytes themselves release NO via an iNOS dependent pathway [37].

The mechanisms by which NO causes cell injury are manifold; it complexes with non-heme iron containing enzymes [38] and consequently interferes with cellular mitochondrial respiration, DNA replication, and the Krebs cycle [39, 40]. In combination with superoxide anion, NO forms peroxynitrite, a lipid peroxidizing radical disrupting cell membranes. These mechanisms of toxic action might also be relevant to rodent macrophages, which are endowed with iNOS and hence are capable upon appropriate stimulation to generate NO metabolites. In the rat model of Borna disease, an encephalomyelitis of horses and sheep caused by an RNA virus, Zheng et al. demonstrated the presence of macrophages within the inflammatory lesion within the brain, however, the number of iNOS positive cells was much lower compared to the total number of macrophages localized [41]. The low percentage of iNOS positive cells may be the result of different stages of activation within this cell population or could point to iNOS expression as an early and transient phenomenon during the course of neuroinflammation.

In addition to the neurotoxic effects, reactive nitrogen intermediates are likely to affect the BBB and BNB, respectively, contributing to increased vasodilatation and thus promoting edema formation.

Conversely, NO may also play a beneficial role in inflammatory diseases of the nervous system. As previously mentioned, NO does regulate T cell proliferation [42]. Our group demonstrated a dose-dependent upregulation of iNOS mRNA and precipitation of nitrite upon stimulation with IFNγ and TNFα in rat Schwann cells [43]. This NO production may account, at least in part, for suppressive effects on T cell activation by thymic antigen-presenting cells seen in coculture models in which Schwann cells were present. These findings implicate that Schwann cells may modulate inflammatory reactions in the nervous system by releasing NO, expanding their functional role beyond antigen presentation.

Present data demonstrate that NO produced by NOS, might play an important role in the pathogenesis of inflammatory disorders of the nervous system. Future research should focus on the development of highly specific inhibitors for iNOS that leave cNOS unaffected, to clarify the exact mechanisms of action through which this inducible enzyme is involved during neuroinflammation.

Matrix metalloproteinases

Structure and regulation

The matrix metalloproteinases (MMPs) or matrixins comprise a large subfamily of Zn^{2+}-dependent neutral endoproteinases, which includes the collagenases, gelatinases, stromelysins, matrilysin, metalloelastase, and membrane-type metalloproteinases, each of which is the product of a different gene. They all have a catalytic zinc-binding domain in common that includes a sequence motif HEXXH in which Glu (E) acts as a catalytic base [44]. Whereas the recently discovered membrane-type MMPs are bound to the cellular surface, all other MMPs are secreted into the extracellular space by a wide range of cell-types as latent pro-enzymes requiring activation by proteolytic cleavage of an amino-terminal domain to expose the active catalytic site [45]. Various cells have been demonstrated to be capable of expressing different MMPs, including activated T lymphocytes [46], macrophages [47], astrocytes, and microglial cells [48, 49].

The regulation of MMPs is strictly controlled at various stages. At the transcriptional level, different cytokines, such as TNFα, IL-1, TGFβ, and eicosanoids, such as prostaglandin E_2 (PGE_2), can directly induce or suppress MMP expression [50]. Since the same cytokine can exert a stimulating effect on one cell and a suppressive effect on another in respect to the expression of the same MMP, this regulatory activity seems more likely to be dependent on the MMP signal transduction/synthesis cascade within the targeted cell, rather than on the cytokine acting as a ligand. Once synthesized MMPs are secreted as inactive zymogens, they are cleaved in various steps to their final active form [51]. This process of activation is regulated by MMPs and other proteinases, such as plasmin. The active forms of MMPs are subject to inhibition by the tissue inhibitors of the MMP family (TIMPs), which are expressed ubiquitously in the extracellular milieu and form a complex of 1:1 stoichiometry with the endoproteinases [52].

MMPs are capable of degrading all protein components of the extracellular matrix (ECM). Their substrate specificities are broad but not overlapping. ECM degradation is an important mechanism in various physiologic as well as pathologic processes, in which the finely tuned regulation of MMPs appears to be of critical

importance. Any increase in enzymatic activity will consequently result in cell invasion and tissue destruction, whereas an imbalance in favor of inhibitory mechanisms
will lead to fibrosis [45]. Proteins of the ECM are, however, not the only targets of
enzymatic activity of MMPs. Recent studies revealed that also inactive proforms of
MMPs [53], enzyme inhibitors such as α1-antitrypsin [54], cell adhesion molecules
such as L-selectin [55], cytokine precursors, and cytokine receptors are substrates
for various MMPs. The shedding of cytokine receptors, such as the 80 kDa TNF
receptor, the interleukin-6 receptor α-subunit, and the cytokine proforms of TGFα
and TNFα, were prevented by synthetic pseudo-peptide MMP-inhibitors [56-58].
However, the TNFα converting enzyme (TACE) has recently been described to be a
unique disintegrin metalloproteinase with notable sequence identity to the ada-
malysin family of metalloproteinases, implying TACE to more likely be a novel metalloproteinase, distinct from the hitherto known 'classical' MMPs [59, 60]. Further
avenues of research should elucidate the role of MMPs in the release of surface
cytokines and receptors, which impacts markedly on the orchestration of immunin-
flammatory responses.

The role of matrix metalloproteinases in the inflamed nervous system

Substantial evidence has been accumulated so far pointing to MMPs as key enzymes
in the pathogenesis of inflammatory disorders of the nervous system. Recent studies in EAE shed light on their role in pathophysiologic mechanisms involved in
inflammatory demyelination of the CNS.

The cerebrospinal fluid (CSF) of rodents with clinically active EAE and experimental meningitis, respectively, revealed increased levels of 92 kDa gelatinase
(MMP-9) [61]. In EAE, this MMP could be localized to infiltrating mononuclear
cells and to the perivascular area by immunohistochemistry [62]. It is thought, based
on in vitro studies, that MMP-9 facilitates transbasement membrane migration of T
cells [46]. Since migration of autoreactive T cells from blood to the CNS is of paramount importance in the genesis of inflammatory demyelination, such an action
would assign MMPs a strategic role in this process. In fact, Rosenberg and co-workers demonstrated that intracerebral injection or induction of 72 kDa and 92 kDa
gelatinase, respectively, resulted in opening of the BBB and breakdown of the ECM
in rodents [63, 64]. The application of broad spectrum MMP inhibitors suppressed
and reversed clinical EAE in a dose-dependent way [65, 66]. These findings were
paralleled by restoration of the damaged BBB and a significant reduction of MMP-
9 activity in the CNS in treated animals. Recent studies investigating the temporospatial expression pattern of various MMPs during the clinical course of EAE
found matrilysin and the 92 kDa gelatinase to be selectively upregulated during the
course of the disease, pointing to these two MMPs as key candidate enzymes in the
pathogenesis of inflammation within the CNS [62, 67]. In MS-patients, increased

CSF levels of MMP-9 were found to be associated with a leaky BBB as demonstrated by gadolinium-enhanced magnetic resonance imaging (MRI). In this patient group, treatment with high-dose methylprednisolone, an MMP transcription inhibitor, reduced both contrast on brain MRI and CSF levels for MMP-9, interpreted as a result of either reduced MMP expression, or increased TIMP production [68]. Both situations, consequently, lead to increased enzymatic activity, disrupting the basal lamina around capillaries, and paving the way for inflammatory cells and other soluble substances into the CNS.

In an attempt to localize MMPs within the inflamed CNS, several investigations have been performed: In demyelinating MS lesions MMP-9 was localized to macrophages and astrocytes. Astrocytes also expressed this MMP in chronic lesions. Within the MS white matter, mononuclear cells in perivascular cuffs were found to be positive for MMP-9 [48, 49]. Cytokines documented as inducing transcription of the 92 kDa gelatinase were also found within active MS lesions both in perivascular cells and in activated microglia [69]. Other MMPs were also detected in MS plaques and macrophages exhibited interstitial collagenase, 72 kDa gelatinase, and stromelysin-1. The latter two were also localized to astrocytes [49]. However, detection and localization of MMPs within the MS lesion does not provide entire information about the proteolytic activity of the enzyme itself. Moreover, since inhibition of the activated forms of MMPs by TIMPs and other proteinase inhibitors plays a crucial role in MMP regulation, further studies are needed to elucidate the function of these enzymes.

This accumulated evidence sheds light on mechanisms by which MMPs could contribute to the pathogenesis and progression of inflammatory demyelinating diseases of the CNS. They are instrumental in mediating T-cell migration, disrupt the BBB, and release the myelinotoxic cytokine TNFα. The myelin sheath could be another target of MMP action, since these enzymes are known to degrade myelin basic protein (MBP), one of its major protein components [70]. Thus, MMPs could contribute to demyelination and perpetuate the inflammatory response by generating additional immunogenic peptides.

The role of MMPs in the PNS has not been the object of intensive research, and few studies have investigated this question. Similar to investigations in the CNS, broad spectrum MMP inhibitors were applied to EAN models. Redford et al. [71] significantly ameliorated the clinical course of EAN in rodents, implicating a potential role of MMPs in the pathogenesis of inflammatory disorders of the PNS. Our group investigated the temporospatial expression pattern of various MMPs during the clinical course of EAN. As in the CNS, MMPs do not respond to inflammatory stimuli in an all-or-none fashion, but rather are differentially regulated in the PNS. Matrilysin and the 92 kDa gelatinase were found to be significantly upregulated during the course of EAN. The same two MMPs were found to be present in sural nerve biopsies from patients with GBS, but not in non-inflammatory controls, underlining the data obtained from the animal model [72]. However, differences in quantitative amounts as well as in the temporospatial expression of MMPs can be seen comparing EAE and EAN, pointing to different mechanisms of the response to inflammation in the CNS and PNS. Regulation of MMP activity appears more likely to be a system-specific answer to an inflammatory stimulus and thus MMPs may contribute to the pathogenesis of inflammatory demyelinating disorders of the CNS and PNS in different ways. Clearly, further research is required to clarify the involvement of MMPs in this process.

Other proteases

The idea that proteinases might be involved in demyelinating disease of the nervous system dates back to the late 1960s [73]. Broad spectrum inhibitors of proteinase activity, especially inhibiting plasminogen activator and other neutral proteinases, conferred significant protection against clinical expression of EAE in Lewis rats [74]. Using the same inhibitor in EAN, a delay in the development of clinical signs was seen, whereas clinical disease severity remained unchanged [75]. Interpreting these data retrospectively, one has to consider that these broad spectrum inhibitors, at least partially, also inhibited MMPs. Plasmin is a key proteinase catalyzing the activation of various MMPs by proteolytic cleavage [76]. On the other hand, it is widely accepted that macrophages contain and secrete a large variety of lysosomal neutral and acid proteases. Increased levels of the latter have been described in the PNS during the course of EAN [77]. But also other cell types, such as mast cells [78] and astrocytes [79] have been shown to express different proteases. Once released these proteases could act by exhibiting direct cellular toxicity as well as by regulating immunoactive mediators, such as cytokines and MMPs.

Since currently no data are available regarding the mode of activation of each individual protease *in vivo* the contributory effect to the inflammatory process within the nervous system can only be assumed.

Cyclooxygenase

Cyclooxygenase (COX) is a rate-limiting enzyme catalyzing the synthesis of prostanoids from arachidonic acid [80]. Two isoforms of the membrane protein COX have been described in mammalian cells so far: (a) COX-1, which is constitutively expressed and responsible for the physiologic production of prostaglandins; and (b) COX-2, a highly inducible enzyme, under strict regulation of different cytokines, mitogens, reactive oxygen intermediates, and endotoxins, which is responsible for the increased production of prostaglandins during inflammation [81–83]. Moreover, NO may have a direct effect on activating COX resulting in the overproduction of the inflammatory prostaglandin PGE_2 [84] and PGF_2 [85] in microglial cells [86] and astrocytes [87], respectively.

Whereas COX-1 seems to be ubiquitously expressed in most tissues and organs, COX-2 is primarily localized to inflammatory cells and tissues. Astrocytes, microglial, and Schwann cells in rodents are both known to produce prostanoids via a COX pathway in response to e.g. phorbol diester, bacterial lipopolysaccharide, the neuropeptide Substance P or the phlogogenic cytokine IL-1β [88–93]. Both isoforms of COX have also been reported to be present in human monocytes [94].

The monocyte-macrophage system is an important effector in the pathogenesis of immune mediated inflammation in the nervous system. First of all, macrophages physically impart damage to the myelin sheath through phagocytosis. Secondly, they elaborate an array of proinflammatory and cytotoxic molecules that are released into the local microenvironment. Besides their capacity to produce NO, macrophages can generate prostaglandins upon activation. Several studies found a positive correlation between an increase in PGE_2 and the clinical severity of EAE [95, 96]. Consequently, the role of prostaglandins and COX in inflammatory diseases of the CNS have been investigated further. Clinical disease severity in rodent EAE could be ameliorated by both the application of neutralizing antibodies to PGE and COX-inhibitors [97–100], pointing to an important role of COX in the pathogenesis of autoimmune inflammation within the CNS.

Information about the role of COX in the inflamed PNS is limited. An increase in mRNA COX-2 transcripts within the spinal cord has been reported following peripheral inflammation [101], and administration of COX inhibitors (and, parenthetically, joint cyclooxygenase and lipoxygenase blockers), indomethacin and BW 755c, abrogated the clinical, neurophysiological, and histomorphological expression of Lewis rat EAN induced with whole peripheral myelin [13].

In summary, there is a body of evidence emphasizing the role of COX in neuroinflammation. However, the relative contribution of the inducible form COX-2 and the constitutively expressed form COX-1 to the inflammatory response within the nervous system remains unclear. Further studies, by applying selective inhibitors [102], should allow differentiation between the individual action of each subform of COX.

Conclusion

There is an emerging body of evidence implicating a central role of inducible enzymes in the inflammatory immune response of the nervous system. Collective data suggests that these enzymes might contribute to the process of inflammation as well as demyelination. Functionally, they seem to be involved in the recruitment of inflammatory cells, in damaging the BBB and BNB, respectively, in axonal and neuronal toxicity as well as myelin damage, and in modulating the inflammatory mechanism. These multifarious actions are part of the complexity of the immune response within the nervous system.

It is hoped that the exponential progress of research in cellular and humoral immunity during the latter years will increase our knowledge about the enzymatic cascades involved in inflammatory reactions. Based on this growing understanding, new synthetic modulators of inducible enzymatic activity should be designed which hopefully will enlarge our therapeutic armentarium for the treatment of inflammatory diseases of the nervous system in the future.

References

1   Swanborg RH (1995) Experimental autoimmune encephalomyelitis in rodents as a model for human demyelinating disease. *Clin Immunol Immunopathol* 77: 4–13

2   Wekerle H, Kojima K, Lannes-Vieira J, Lassmann H, Linington C (1994) Animal models. *Ann Neurol* 36: S47–S53

3   Wekerle H (1993) Experimental autoimmune encephalomyelitis as a model of immune-mediated CNS disease. *Curr Opin Neurobiol* 3: 779–784

4   Lassmann H, Zimprich F, Rössler K, Vass K (1991) Inflammation in the nervous system. Basic mechanisms and immunological concepts. *Rev Neurol (Paris)* 147: 763–781

5   Raine CS (1992) Demyelinating diseases. In: RL Davis, DM Robertson (eds): *Textbook of neuropathology*. Williams and Wilkins, Baltimore, 535–552

6   Hartung H-P, Pollard JD, Harvey GK, Toyka KV (1995) Invited review – immunopathogenesis and treatment of the Guillain-Barré syndrome, Part I and II. *Muscle Nerve* 18: 137–164

7   Hartung H-P, Heininger K, Schäfer B, Fierz W, Toyka KV (1988) Immune mechanisms in inflammatory polyneuropathy. *Ann NY Acad Sci* 545: 122–161

8   Powell HC, Olee T, Brostoff SW, Mizisin AP (1991) Comparative histology of experimental allergic neuritis induced with minimum length neuritogenic peptides by adoptive transfer with sensitized cells or direct sensitization. *J Neuropathol Exp Neurol* 50: 658–674

9   Arnason BGW, Soliven B (1993) Acute inflammatory demyelinating polyradiculoneuropathy. In: PJ Dyck, PK Thomas, JW Griffin, PA Low, J Poduslo (eds): *Peripheral neuropathy*. Saunders, Philadelphia, 1437–1497

10 Hartung H-P, Stoll G, Toyka KV (1993) Immune reactions in the peripheral nervous system. In: PJ Dyck, PK Thomas, JW Griffin, PA Low, J Poduslo (eds): Peripheral Neuropathies. Saunders, Philadelphia, 418–444

11 Hartung H-P (1995) Pathogenesis of inflammatory demyelination: implications for therapy. *Curr Opin Neurol* 8: 191–199

12 Bianchi E, Bender JR, Blasi F, Pardi R (1997) Through and beyond the wall: late steps in leukocytes transendothelial migration. *Immunol Today* 18: 586–591

13 Hartung HP, Schäfer B, Heininger K, Stoll G, Toyka KV (1988) The role of macrophages

and eicosanoids in the pathogenesis of experimental allergic neuritis. Serial clinical, electrophysiological, biochemical and morphological observations. *Brain* 111: 1039–1059

14   Moncada S, Higgs A (1993) The L-arginine-nitric oxide pathway. *New Engl J Med* 329: 2002–2012

15   Stuehr DJ, Nathan CF (1989) Nitric oxide. A macrophage product responsible for cytostasis and respiratory inhibition in tumour target cells. *J Exp Med* 169: 1543–1555

16   Kirk SJ, Regan MC, Barbul A (1990) Cloned murine T lymphocytes synthesize a molecule with the biological characteristics of nitric oxide. *Biochem Biophys Res Commun* 173: 660–665

17   McCall TB, Boughton-Smith NK, Palmer RMJ, Whittle BJR, Moncada S (1989) Synthesis of nitric oxide from L-arginine by neutrophils. *Biochem J* 261: 293–296

18   Curran RD, Billiar TR, Stuehr DJ, Hofmann K, Simmons RL (1989) Hepatocytes produce nitrogen oxides from L-arginine in response to inflammatory stimuli. *J Exp Med* 170: 1769–1774

19   Werner-Felmayer G, Werner ER, Fuchs D, Hausen A, Reibnegger G, Wachter H (1990) Tetrahydrobiopterin-dependent formation of nitrite and nitrate in murine fibroblasts. *J Exp Med* 172: 1599–1607

20   Oswald IP, Gazzinelli RT, Sher A, James SL (1992) IL-10 synergizes with IL-4 and transforming growth factor-β to inhibit macrophage cytotoxic activity. *J Immunol* 148: 3578–3582

21   McCall TB, Palmer RM, Moncada S (1991) Interleukin-8 inhibits the induction of nitric oxide synthase in rat peritoneal neutrophils. *Biochem Biophys Res Commun* 186: 680–685

22   Sicher SC, Vazquez MA, Lu CY (1994) Inhibition of macrophage Ia expression by nitric oxide. *J Immunol* 153: 1293–1300

23   Tayler-Robinson AW, Liew FY, Severn A, Xu D, McSorley S, Garside P, Padron J, Phillips RS (1994) Regulation of the immune response by nitric oxide differentially produced by T-helper type-1 and T-helper type-2 cells. *Eur J Immunol* 24: 980–984

24   Rengasamy A, Johns RA (1993) Regulation of nitric oxide synthase by nitric oxide. *Mol Pharmacol* 44: 124–128

25   Radomski MW, Palmer RMJ, Moncada S (1990) Glucocorticoids inhibit the expression of an inducible, but not constitutive, nitric oxide synthase in vascular endothelial cells. *Proc Natl Acad Sci USA* 87: 10043–10047

26   Parkinson JF, Mitrovic B, Merill JE (1997) The role of nitric oxide in multiple sclerosis. *J Mol Med* 75: 174–186

27   Koprowski H, Zheng YM, Heber-Katz E, Fraser N, Rorke L, Fu ZF, Hanlon C, Dietzschold B (1993) *In vivo* expression of inducible nitric oxide synthase in experimentally induced neurologic disease. *Proc Natl Acad Sci USA* 90: 3024–3027

28   MacMicking JD, Willenborg DO, Weidemann MJ, Rocket KA, Cowden WB (1992) Elevated secretion of reactive nitrogen and oxygen intermediates by inflammatory leukocytes in hyperacute experimental autoimmune encephalomyelitis: enhancement by the soluble products of encephalitogenic T cells. *J Exp Med* 176: 303–307

29   Bö L, Dawson TM, Wesselingh S, Sverre M, Choi S, Kong PA, Hanley D, Trapp BD (1994) Induction of nitric oxide synthase in demyelinating regions of multiple sclerosis brains. *Ann Neurol* 36: 778–786

30   Lin RF, Lin TS, Tilton RG, Cross AH (1993) Nitric oxide localized to spinal cord of mice with experimental allergic encephalomyelitis: an electron paramagnetic study. *J Exp Med* 178: 643–648

31   Cross AH, Misko TP, Lin RF, Hickey WF, Trotter JL, Tilton RF (1994) Aminoguanidine, an inhibitor of inducible nitric oxide synthase, ameliorates experimental autoimmune encephalomyelitis in SJL mice. *J Clin Invest* 93: 2684–2690

32   Zhao W, Tilton RG, Corbett JA, McDaniel ML, Misko TP, Williamson JR, Cross AH, Hickey WF (1996) Experimental allergic encephalomyelitis in the rat is inhibited by aminoguanidine, an inhibitor of nitric oxide synthase. *J Neuroimmunol* 64: 123–133

33   Brenner T, Brocke S, Szafer F, Sobel RA, Parkinson JF, Perez DH, Steinmann L (1997) Inhibition of nitric oxide synthase for treatment of experimental autoimmune encephalomyelitis. *J Immunol* 158: 2940–2946

34   Zielasek J, Jung S, Gold R, Liew FY, Toyka KV, Hartung H-P (1995) Administration of nitric oxide synthase inhibitors in experimental autoimmune neuritis and experimental autoimmune encephalomyelitis. *J Neuroimmunol* 58: 81–88

35   Ruuls SR, van der Linden S, Sontrop K, Huitinga I, Dijkstra C (1996) Aggravation of experimental allergic encephalomyelitis (EAE) by administration of nitirc oxide (NO) synthase inhibitors. *Clin Exp Immunol* 103: 467–474

36   Merrill JE, Ignarro LJ, Sherman MP, Melinek J, Lane TE (1993) Microglial cell cytotoxicity of oligodendrocytes is mediated through nitric oxide. *J Immunol* 151: 2132–2141

37   Merrill JE, Murphy SP, Mitrovic B, Mackenzie-Graham A, Dopp JC, Ding MZ, Griscavage J, Ignarro LJ, Lowenstein CJ (1997) Inducible nitric oxide synthase and nitric oxide production by oligodendrocytes. *J Neuroscie Res* 48: 372–384

38   Pellat C, Henry Y, Drapier J-C (1990) IFNγ activated macropahges: detection of complexes between L-arginine-derived nitric oxide and non-heme-iron protein. *Biophys Res Commun* 166: 119–125

39   Kolb H, Kolb-Bachofen V (1992) Nitric oxide: a pathogenic factor in autoimmunity. *Immunol Today* 13: 157–160

40   Zhang J, Dawson VL, Dawson TM, Snyder SH (1994) Nitric oxide activation of poly(ADP-ribose) synthetase in neurotoxicity. *Science* 263: 687–689

41   Zheng YM, Schafer MK-H, Weihe E, Sheng H, Corisdeo S, Fu ZF, Koprowski H, Dietzschold B (1993) Severity of neurological signs and degree of inflammatory lesions in the brains of rats with Borna diseases correlate with the induction of nitric oxide synthase. *J Virol* 67: 5786–5791

42   Liew FY (1995) Regulation of lymphocyte functions by nitric oxide. *Curr Opin Immunol* 7: 396–399

43   Gold R, Zielasek J, Kiefer R, Toyka KV, Hartung H-P (1996) Secretion of nitrite by Schwann cells and its effect on T-cell activation *in vitro*. *Cell Immunol* 168: 69–77

44 Woessner Jr JF (1994) The family of matrix metalloproteinases. *Ann NY Acad Sci* 732: 11–21

45 Birkedal-Hansen H (1995) Proteolytic remodeling of extracellular matrix. *Curr Opin Cell Biol* 7: 728–735

46 Leppert D, Waubant E, Galardy R, Bunnett NW, Hauser SL (1995) T cell gelatinases mediate basement membrane transmigration *in vitro*. *J Immunol* 154: 4379–4389

47 Nielsen BS, Timshel S, Kjeldsen L, Sehested M, Pyke C, Borregaard N, Dano K (1996) 92kDa type IV collagenase (MMP-9) is expressed in neutrophils and macrophages but not in malignant epithelial cells in human colon cancer. *Int J Cancer* 65: 57–62

48 Cuzner ML, Gveric D, Strand C, Loughlin AJ, Paemen L, Opdenakker G, Newcombe J (1996) The expression of tissue-type plasminogen activator, matrix metalloproteases and endogenous inhibitors in the central nervous system in multiple sclerosis: comparison of stages in lesion evolution. *J Neuropathol Exp Neurol* 55: 1194–1204

49 Maeda A, Sobel RA (1996) Matrix metalloproteinases in the normal human central nervous system, microglial nodules, and multiple sclerosis lesions. *J Neuropathol Exp Neurol* 55: 300–309

50 Ries C, Petrides PE (1995) Cytokine regulation of matrix metalloproteinase activity and its regulatory dysfunction in disease. *Biol Chem Hoppe-Seyler* 376: 345–355

51 Murphy G, Willenbrock F, Crabbe T, O'Shea M, Ward R, Atkinson S, O'Connell J, Docherty A (1994) Regulation of matrix metalloproteinase activity. *Ann NY Acad Sci* 732: 31–41

52 Overall CM (1994) Regulation of tissue inhibitor of matrix metalloproteinase expression. *Ann NY Acad Sci* 732: 51–64

53 Vassalli J-D, Pepper MS (1994) Membrane proteases in focus. *Nature* 370: 14–15

54 Sires UI, Murphy G, Baragi VM, Fliszar CJ, Welgus HG, Senior RM (1994) Matrilysin is much more efficient than other matrix metalloproteinases in the proteolytic inactivation of α1-antitrypsin. *Biochem Biophys Res Commun* 204: 613–620

55 Preece G, Murphy G, Ager A (1996) Metalloproteinase-mediated regulation of L-selectin levels on leucocytes. *J Biol Chem* 271: 11634–11640

56 Crowe PD, Walter BN, Mohler KM, Otto-Evans C, Black RA, Ware CF (1995) A metalloproteinase inhibitor blocks shedding of the 80 kDa TNF receptor and TNF processing in T lymphocytes. *J Exp Med* 181: 1205–1210

57 Gearing AJH, Beckett P, Christodoulou M, Churchill M, Clements J, Davidson AH, Drummond AH, Galloway WA, Gilbert R, Gordon JL et al (1994) Processing of tumor necrosis factor-α precursor by metalloproteinases. *Nature* 370: 555–557

58 Mohler KM, Sleath PR, Fitzner JN, Cerretti DP, Alderson M, Kerwar SS, Torrance DS, Otten-Evans C, Greenstreet T, Weerawarna K et al (1994) Protection against lethal dose of endotoxin by an inhibitor of tumor necrosis factor processing. *Nature* 370: 218–220

59 Black RA, Rauch CT, Kozlosky CJ, Peschon JJ, Slack JL, Wolfson MF, Castner BJ, Stocking KL, Reddy P, Srinivasan S et al (1997) A metalloproteinase disintegrin that releases tumor-necrosis factor-α from cells. *Nature* 385: 729–733

60 Moss ML, Jin S-LC, Milla ME, Burkhart W, Carter HL, Chen W-J, Clay WC, Didsbury

JR, Hassler D, Hoffmann CR et al (1997) Cloning of a disintegrin metalloproteinase that processes precursor tumor necrosis factor-α. *Nature* 385: 733–736

61 Gijbels K, Proost P, Carton H, Billiau A, Opdenakker G (1993) Gelatinase B is present in the cerebrospinal fluid during experimental autoimmune encephalomyelitis and cleaves myelin basic protein. *J Neurosci Res* 36: 432–440

62 Kieseier BC, Kiefer R, Clements JM, Miller K, Wells GMA, Schweitzer T, Gearing AJH, Hartung H-P (1998) Matrix metalloproteinase-9 and -7 are regulated in experimental autoimmune encephalomyelitis. *Brain* 121: 159–166

63 Rosenberg GA, Kornfeld M, Estrada E, Kelley RO, Liotta LA, Stetler-Stevenson WG (1992) TIMP-2 reduces proteolytic opening of blood-brain barrier by type IV collagenase. *Brain Res* 576: 203–207

64 Rosenberg GA, Dencoff JE, McGuire PG, Liotta LA, Stetler-Stevenson WA (1994) Injury-induced 92-kilodalton gelatinase and urokinase expression in rat brain. *Lab Invest* 71: 417–422

65 Gijbels K, Galardy RE, Steinman L (1994) Reversal of experimental autoimmune encephalomyelitis with a hydroxamate inhibitor of matrix metalloproteinases. *J Clin Invest* 94: 2177–2182

66 Hewson AK, Smith T, Leonard JP, Cuzner ML (1995) Suppression of experimental allergic encephalomyelitis in the Lewis rat by the matrix metalloproteinase inhibitor Ro31-9790. *Inflamm Res* 44: 345–349

67 Clements JM, Cossins JA, Wells GMA, Corkill DJ, Helfrich K, Wood LM, Pigott R, Stabler G, Ward GA, Gearing AJH et al (1997) Matrix metalloproteinase expression during experimental autoimmune encephalomyelitis and effects of a combined matrix metalloproteinase and tumor necrosis factor-α inhibitor. *J Neuroimmunol* 74: 85–94

68 Rosenberg GA, Dencoff JE, Correa N, Reiners M, Ford CC (1996) Effect of steroids on CSF matrix metalloproteinases in multiple sclerosis: relation to blood-brain barrier injury. *Neurology* 46: 1626–1632

69 Canella B, Raine SC (1995) The adhesion molecule and cytokine profile of multiple sclerosis lesions. *Ann Neurol* 37: 424–435

70 Chandler S, Coates R, Gearing A, Lury J, Wells G, Bone E (1995) Matrix metalloproteinases degrade myelin basic protein. *Neurosci Lett* 201: 223–226

71 Redford EJ, Smith KJ, Gregson NA, Davies M, Hughes P, Gearing AJH, Miller K, Hughes RAC (1997) A combined inhibitor of matrix metalloproteinase activity and tumour necrosis factor-alpha processing attenuates experimental autoimmune neuritis. Brain 120: 1895–1905

72 Kieseier BC, Clements JM, Pischel HB, Wells GMA, Miller K, Gearing AJH, Hartung H-P (1998) Matrix metalloproteinases MMP-9 and MMP-7 are expressed in experimental autoimmune neuritis and the Guillain-Barré Syndrome. *Ann Neurol* 43: 427–434

73 Rinne UK, Riekkinen P (1968) Esterase, peptidase and proteinase activities of human cerebrospinal fluid in multiple sclerosis. *Acta Neurol Scand* 44: 156–167

74 Brosnan CF, Cammer W, Norton WT, Bloom BR (1980) Proteinase inhibitors suppress the development of experimental encephalomyelitis. *Nature* 285: 235–237

75 Schabet M, Whitaker JN, Schott K, Stevens A, Zürn A, Bühler R, Wiethölter H (1991) The use of protease inhibitors in experimental allergic neuritis. *J Neuroimmunol* 31: 265–272

76 Krane SM (1994) Clinical importance of matrix metalloproteinases and their inhibitors. *Ann NY Acad Sci* 732: 1–10

77 Sobue G, Yamoto S, Matsuoka M, Uematsu J, Sobue I (1982) The role of macrophages in demyelination in experimental allergic neuritis. *J Neurol Sci* 44: 229–239

78 Rouleau A, Dimitriadou V, Trung-Tuong MD, Newlands GF, Miller HR, Schwartz JC, Garbarg M (1997) Mast cell specific proteases in rat brain: changes in rats with experimental allergic encephalomyelitis. *J Neural Transm* 104: 399–417

79 Mucke L, Eddleston M (1993) Astrocytes in infectious and immune-mediated diseases of the central nervous system. *FASEB J* 7: 1226–1232

80 Smith WL (1989) The eicosanoids and their biochemical mechanisms of action. *Biochem J* 259: 315–324

81 Porreca E, Reale M, Di Febbo C, Di Gioacchino M, Barbacane RC, Castellani ML, Baccante G, Conti P, Cuccurullo F (1996) Down-regulation of cyclooxygenase-2 (COX-2) by interleukin-1 receptor antagonist in human monocytes. *Immunol* 89: 424–429

82 Feng L, Xia Y, Garcia GE, Hwang D, Wilson CB (1995) Involvement of reactive oxygen intermediates in cyclooxygenase-2 expression induced by interleukin-1, tumor necrosis factor-α and lipopolysaccharide. *J Clin Invest* 95: 1669–1675

83 Cao C, Matsumura K, Watanabe Y (1997) Induction of cyclooxygenase-2 in the rat brain by cytokines. *Ann NY Acad Sci* 813: 307–309

84 Corbett JA, Kwon G, Turk J, McDaniel ML (1993) IL-1 beta induces the coexpression of both nitric oxide synthase and cyclooxygenase by islets of Langerhans: activation of cyclooxygenase by nitric oxide. *Biochemistry* 32: 13767–13770

85 Janabi N, Chabrier S, Tardieu M (1996) Endogenous nitric oxide activates prostaglandin F2 alpha production in human microglia cells but not in astrocytes: a study of interactions between eicosanoids, nitric oxide, and superoxide anion (O2-) regulatory pathways. *J Immunol* 157: 2129–2135

86 Guastadisegni C, Minghetti L, Nicolini A, Polazzi E, Ade P, Balduzzi M, Levi G (1997) Prostaglandin E2 synthesis is differentially affected by reactive nitrogen intermediates in cultured rat microglia and RAW 264.7 cells. *FEBS Lett* 413: 314–318

87 Molina-Holgado F, Lledo A, Guaza C (1995) Evidence for cyclooxygenase activation by nitric oxide in astrocytes. *Glia* 15: 167–172

88 Constable AL, Armati PJ, Toyka KV, Hartung H-P (1994) Production of prostanoids by Lewis rat Schwann cells *in vitro*. *Brain Res* 635: 75–80

89 Hartung H-P, Toyka KV (1987) Leukotriene production by cultured astroglial cells. *Brain Res* 435: 367–370

90 Hartung H-P, Toyka KV (1987) Phorbol diester TPA elicits prostaglandin E release from cultured rat astrocytes. *Brain Res* 417: 347–349

91 Hartung H-P, Schäfer B, Heininger K, Toyka KV (1989) Recombinant interleukin-1 beta

stimulates eicosanoid production in rat primary culture astrocytes. *Brain Res* 489: 113–119

92   Hartung H-P, Heininger K, Schäfer B, Toyka KV (1988) Substance P and astrocytes: stimulation of the cyclooxygenase pathway of arachidonic acid metabolism. *FASEB J* 2: 48–51

93   Slepko N, Minghetti L, Polazzi N, A, Levi G (1997) Reorientation of prostanoid accompanies "activation" of adult microglia cells in culture. *J Neurosci Res* 49: 292–300

94   Fu J-Y, Masferrer J-L, Seibert N, Raz A, Needleman P (1990) The induction and suppression of prostaglandin H2 synthase (cyclooxygenase) in human monocytes. *J Biol Chem* 265: 16737–16740

95   Bolton C, Gordon D, Turk JL (1984) A longitudinal study of the prostaglandin content of the central nervous system tissues from guinea pigs with acute EAE. *Int J Immunopharmacol* 6: 155–161

96   Fretland DJ (1992) Potential role of prostaglandins and leukotrienes in multiple sclerosis and experimental allergic encephalomyelitis. *Prostaglandins Leukot Essent Fatty Acids* 45: 249–257

97   Mertin J, Stackpoole A (1981) Anti-PGE antibodies inhibit *in vivo* development of cell-mediated immunity. *Nature* 194: 456–458

98   Weber F, Meyermann R, Hempel K (1991) Experimental allergic encephalomyelitis - prophylactic and therapeutic treatment with the cyclooxygenase inhibitor piroxicam (Feldene). *Int Arch Allergy Appl Immunol* 95: 136–141

99   Simmons RD, Hugh AR, Willenborg DO, Cowden WB (1992) Suppression of the active but not passive autoimmune encephalomyelitis by dual cyclo-oxygenase and 5-lipoxygenase inhibition. *Acta Neurol Scand* 85: 949–954

100 Reder AT, Thapar M, Sapugay AM, Jensen MA (1994) Prostaglandins and inhibitors of arachidonate metabolism suppress experimental allergic encephalomyelitis. *J Neuroimmunol* 54: 117–127

101 Beiche F, Scheuerer S, Brune K, Geisslinger G, Goppelt-Struebe M (1996) Up-regulation of cyclooxygenase-2 mRNA in the rat spinal cord following peripheral inflammation. *FEBS Lett* 390: 165–169

102 Kurumbail RG, Stevens AM, Gierse JK, McDonald JJ, Stegeman RA, Pak JY, Gildehaus D, Miyashiro JM, Penning TD, Seibert K et al (1996) Structural basis for selective inhibition of cyclooxygenase-2 by anti-inflammatory agents. *Nature* 384: 644–648

Inducible enzymes in inflammation: advances, interactions and conflicts

Annette Tomlinson and Derek A. Willoughby

Experimental Pathology, St. Bartholomew's and The Royal London School of Medicine and Dentistry, Charterhouse Square, London EC1M 6BQ, UK

Inducible nitric oxide synthase, cyclooxygenase and heme oxygenase

Isoforms of enzymes which are induced in acute and chronic inflammation, have become important as potential therapeutic targets. A great deal of attention has focused on inducible nitric oxide synthase (iNOS), the enzyme which catalyses L-argininine to L-citrulline and nitric oxide (NO), and on the recently discovered second isoform of cyclooxygenase (COX-2), which converts arachidonic acid to bioactive prostanoids.

Current scientific opinion, summarised in the chapter by Hobbs and Moncada in this volume, indicates that NO is an important molecule in inflammation. Physiological production of NO by the endothelium, from constitutively expressed NOS, maintains vascular tone and vessel patency [1]. Nitric oxide contributes to prevent platelet aggregation and leucocyte adhesion to the vessel wall by regulation of adhesion molecule expression. By these actions NO can be said to be *anti*-inflammatory. However, by increasing vasodilatation and vascular permeability, NO from this source is *pro*-inflammatory in the early stages of acute inflammation [2]. Nitric oxide generated from iNOS by inflammatory cells is an important molecule in host defence against bacteria, viruses and intracellular parasites [1]. Inappropriate or excessive production, however, may result in the massive vasodilatation evidenced in septic shock [3]. Moreover, combination with reactive oxygen and nitrogen species may lead to peroxynitrite formation, nitration of protein tyrosine, hydroxyl radical production and tissue damage [4].

In a similar fashion, prostacyclin (PGI_2), the major prostaglandin formed by the endothelium from constitutively expressed COX-1, aids vascular homeostasis, synergising with NO to prevent platelet aggregation [5]. However, PGI_2 and prostaglandin E_2 (PGE_2), produced by the inducible isoform COX-2 from inflammatory cell sources, are early mediators of acute inflammation, characterised by vasodilatation and plasma extravasation [6]. Our study of inducible isoforms of COX and NOS in the murine chronic granulomatous tissue air pouch was the first to have demonstrated that in acute and chronic inflammation, the inflammatory profile correlates with the increase in COX-2 protein expression, whilst COX-1 isoform expression appears unchanged [7]. This has led to COX-2 being considered as the major *pro*-inflammatory isoform (see chapter by Pairet et al., this volume). However, the *anti*-inflammatory immunosuppressive properties of prostaglandins in chronic inflammation, in particular PGE_2, are well documented and must not be ignored [8].

Recently, research in our laboratories has focused on the inducible isoform of the enzyme heme oxygenase (HO-1), which in contrast to iNOS and COX-2, appears to be maximally expressed to coincide with the suppression and resolution of a variety of acute inflammations [9] (for an overview see chapter by Willis, this volume). Heme oxygenase catabolizes heme to biliverdin, which is converted enzymatically to bilirubin, (a potent antioxidant and anticomplement bile pigment), free iron and carbon monoxide (CO). Carbon monoxide in common with NO is a vasodilator. The constitutive isoform of heme oxygenase (HO-2) is expressed in the majority of cell types whilst HO-1, catagorized as a heat shock or stress protein (HSP 32) is induced by stressful stimuli. Stress proteins induced in response to stressful stimuli, afford cytoprotection, possibly by protecting protein conformation and priming cells to survive a second, and otherwise lethal stimulus. By increasing the expression of this endogenously produced protective mechanism, HO-1 has become a potential target for limiting inflammation

Insights from genetically modified animals

Developments in the field of molecular biology have allowed the production of transgenic and gene-deleted mutant animals, which overexpress or are deficient in a particular molecule. The advantage is the evaluation of the importance of that molecule in the intact animal. The outcome, despite the production of heterozygous mutants however, can be an all or nothing response, which does not consider the physiological reaction of the organism to a graded output. Additional compensatory mechanisms may arise from other isoforms of the same molecule or there may be naturally occurring redundancy in the system. With these provisions in mind, animals with genetic disruptions to NOS, COX and HO have provided interesting insights into inflammation.

Deletion of NOS genes

The pharmacological inhibition of NOS to elucidate mechanisms of action of NO in inflammation has suffered from the lack of drug specificity between isoforms. Data from experiments with NOS-deletions were therefore a welcome addition to confirm or refute existing evidence and contribute to our understanding of the inflammatory process. The deletion of the two constitutive NOS genes, endothelial cell NOS (eNOS) and neuronal NOS (nNOS) and that of the inducible NOS have helped clarify the role of NO in inflammation from differing sources.

The eNOS homozygous mutant mouse is described as grossly and histologically normal, with protein absent from the endothelium in a variety of tissues, and as a consequence is hypertensive [10, 11]. Mice lacking eNOS were not protected from the severity of lipopolysaccharide (LPS)-induced endotoxaemia, with mortality rates similar to that of the wild type animal [11]. However, NO produced by endothelium in the brain, protected against focal cerebral ischaemia and reperfusion damage [12]. In contrast, the extent of tissue damage was reduced in the nNOS-deficient mouse [13], as activation of this isoform after ischaemic insult [14] results in glutamate neurotoxicity [15, 16]. In similar studies, deletion of the iNOS gene confirmed the key role of this isoform in the progression of secondary brain tissue damage in the post-ischaemic phase [17, 16].

It has been confirmed by studies in mutant mice that NO is essential for host defence against intracellular parasite and bacterial infection. Mice with a disrupted iNOS gene were highly susceptible to infection with the parasite *Leishmania major* [18]. They survived acute infection with *Toxaplasma gondii*, but NO production was essential at the later stages to prevent parasite expansion [19]. Deletion of the iNOS gene also caused mice to be unable to survive infection with *Trypanosoma cruzi*, the infective agent of Chagas' disease [20]. This inability to mount an effective immune response occurred in the presence of a pro-inflammatory cytokine profile similar to that of wild type controls; interleukin-1 (IL-1) tumour necrosis factor α (TNFα), IL-12, interferon γ (IFNγ) and IL-6. In addition, the progression of *Mycobacterium tuberculosis* infection in mouse lung, accelerated by pharmacological inhibition of iNOS was similar in iNOS-deficient mice [21]. Taken together these data identify NO as a key component in combating infection.

Nitric oxide has also been shown to be an important component of normal wound healing. Closure of excisional wounds was significantly delayed in iNOS-deficient mice, and selective inhibition with iNOS inhibitors showed an identical defect [22]. The delay was completely reversed in mutant mice by a single application of an adenoviral vector containing human iNOS cDNA.

Experimental autoimmune encephalomyelitis (EAE) is a T cell-mediated model of the human demyelinating disease multiple sclerosis. Inflammatory cells, expressing iNOS have been identified in perivascular regions of the central nervous system in both EAE and multiple sclerosis. Pharmacological inhibition of iNOS in EAE gave conflicting results as to the precise role of NO in this model. Often, results depended on the route of drug administration. Inducible NOS-deficient mice, showed greater susceptibility to the onset, exacerbated clinical signs and an inability to reverse the symptoms of EAE in comparison with wild type controls [23]. These data support a protective role for iNOS in this debilitating condition.

A fascinating insight into the variability of impact of NO on autoimmune inflammation has been gained from deletion of the iNOS gene in the MRL-lpr/lpr mouse [24]. The intact MRL-lpr/lpr mouse overexpresses NO and exhibits a variety of inflammatory disorders. Gene deletion allows examination of a varied expression of NO, i.e. markedly decreased in the homozygous knockout in comparison to the wild type and intermediate in the heterozygous mutant. Intriguingly, all three developed glomerulonephritis and synovial pathology, but homozygous mice had less renal vasculitis and IgG rheumatoid factor than wild type controls.

Nitric oxide derived from iNOS has also been identified as a key player in the pathogenesis of intestinal inflammation. Acetic acid-induced colitis in wild type mice increased iNOS mRNA, neutrophil accumulation and tissue edema in the early stages of inflammation, with resolution of the lesion by day 7. Mice deficient in iNOS showed a similar early progression, but at a later stage exhibited massive tissue damage and neutrophil infiltration. These data suggest a protective response to tissue injury, possibly by regulating inflammatory cell trafficking [25]. In LPS-induced gut injury, however, NO *facilitated* the translocation of bacteria to other abdominal tissues [26].

Increasing evidence of the capacity of NO to modulate T cell function and thus the immune response is documented in several chapters of this book. Nitric oxide can reduce T cell proliferation by downregulating MHC class II molecules and IL-2 secretion, and block IFNγ [27, 28]. T cells from iNOS-deficient mice, in response to *in vitro* allogeneic stimulation with BALB/c stimulator cells, increased their proliferative response and produced higher levels of IFNγ, IL-2 and IL-12 when compared to heterozygous controls [29]. Allogeneic skin graft rejection however, was similar in both mutant and wild types. In contrast, heterotopic allogeneic cardiac grafts into iNOS-deficient recipient mice significantly increased perivascular infiltration of mononuclear cells and intimal thickening of vessels and accelerated decline in graft function in comparison to heterologous recipients [30]. These findings support a protective role for iNOS against transplant vasculopathy.

Finally, research on the pathogenesis of LPS-induced septic shock initially suggested a prominent role for iNOS. More recently, evidence supports the involvement of NO in the early haemodynamic effects of endotoxaemia, but iNOS inhibition does not protect from subsequent multiple organ dysfunction and death. Mice lacking the iNOS gene had no survival advantage over wild type controls [31].

On the whole, results from inflammation studies using NOS-deleted animals seem to reaffirm previous findings and appear to be viable experimental tools to add to our understanding. The role of NO in inflammation, however, will depend upon the context in which it is expressed and the levels produced. Further advances will be gained from improved specificity in inhibitors and perhaps more importantly from improved delivery of donors.

Deletion of COX genes

Studies in COX-1- and COX-2-deficient mice provided some surprising outcomes [32, 33]. COX-1 is considered to be the housekeeping gene responsible for basal production of prostaglandins in all tissues and elimination was believed to be responsible for non-steroidal anti-inflammatory (NSAID)-induced gastropathy and renal pathology. The COX-1 knockout mouse, however, appeared healthy and histologically normal except for minor aberrations in renal tubule development. Most unexpected was the lack of lesions in gastric tissues. They also showed less tendency to ulceration than wild type animals after NSAID treatment. Moreover, there was no compensatory production of prostaglandins by COX-2 in the stomach in the absence of COX-1, nor in unstimulated macrophages in response to exogenous arachidonic acid. COX-1 was shown to elaborate basal PGE_2 production in these cells.

Lack of COX-1 resulted in a reduced response to topical administration of arachidonic acid; this was not surprising, given that this isoform can rapidly produce prostaglandins from this exogenous substrate. Ear swelling following application of arachidonic acid was scored at 2 h, an early time point during which COX-1 appears to be important and theoretically insufficient time for the induction of COX-2. Thus, in this model, COX-1 contributed to inflammation. The response to tetradecanoyl phorbol acetate (TPA)-induced ear inflammation scored at 6 h was similar in mutant and wild type animals. It was suggested that this may be due to TPA activating many genes, and not directly affecting COX-1.

COX-2 deficient mice had severe kidney abnormalities and disruption of their reproduction. Their inflammatory response to arachidonic acid-induced ear inflammation, unsurprisingly, did not differ from wild type controls. The response to TPA initially seemed contrary to expectations with no reduction in swelling in comparison to wild type animals. De Witt and Smith evoked the pleiotropic actions of TPA to explain that the loss of a single gene would not diminish its effects [34]. Whilst LPS-stimulated peritoneal macrophages from wild type mice increased their synthesis of PGE_2 10 fold, COX-2-deficient mice produced relatively little prostaglandin showing that COX-2 was inactivated and there was no compensatory induction of COX-1.

Many of the mice developed spontaneous peritonitis, which could not be readily explained. DeWitt and Smith postulated several alternatives: (a) that COX-2 was not important in inflammation; (b) that COX-2 is essential for the prevention of infections before they become life threatening; or (c) that COX-2 is important for the resolution of inflammatory conditions and damaged tissue repair. Since these first publications these issues have not been clarified.

A second study in mice lacking COX-2 supported the original findings [35]. These animals also had diffuse myocardial fibrosis and diminished hepatic toxicity in response to LPS. They also failed to show a suppression of the inflammatory response to arachidonic acid- and TPA- induced ear inflammation and carrageenin-induced paw oedema, despite the fact that many studies had shown the involvement of COX-2 in the latter two inflammatory reactions.

If COX-2 is the major prostaglandin-producing isoform in inflammation then amelioration of inflammatory events would have been expected in COX-2-deleted animals. Perhaps the choice of inflammatory models was not optimal to demonstrate suppression, however lines of evidence suggest that a role for COX-1 in inflammation cannot be totally excluded. Supporting this evidence, in the rat carrageenin-induced pleurisy we have found that COX-2 selective inhibitors were less effective at reducing inflammatory parameters than NSAIDs shown *in vitro* to be more selective for COX-1 [36]. Aspirin and piroxicam inhibited inflammatory cell influx, oedema formation and PGE_2 levels more effectively and over a wider time period of inflammation than selective COX-2 inhibitors. Similar results were observed in the murine chronic granulomatous tissue air pouch, a model of chronic inflammation [37]. We found that NSAIDs more selective for the inhibition of COX-1 were more effective at inhibiting granuloma weight gain, vascularity and COX activity than selective COX-2 inhibitors. These results and others provoked an editorial in the same issue of the journal termed "Is there a COX fight during inflammation?" [38]

Lipopolysaccharide-stimulated peritoneal macrophages from COX-2-deficient mice have been shown to produce very little PGE_2, suggesting that prostaglandins derived from COX-1 are minimal. However in contrast, inhibition of COX-2 protein by dexamethasone in LPS-stimulated peritoneal and alveolar macrophages from intact animals had little effect on PGE_2 levels [39]. It was concluded that in the absence of COX-2, COX-1 could maintain prostaglandin production. In addition, a recent report showed that in immortalized, nontransformed lung fibroblasts, derived from wild type, COX-1 or COX-2-deficient mice, stimulation with IL-1β enhanced expression of the remaining functional gene in both COX-1 and -2-deficient cells. Moreover, cytosolic phospholipase A_2, which regulates substrate availability for COX was also increased. These findings suggest that COX null cells can upregulate the alternate isoform to compensate for deficiencies in prostaglandin synthesis [40].

However, whilst COX-2 inhibitors are highly selective, inhibitors of COX-1 lack specificity and the results described above may also be context and tissue specific. This highlights the anomalies and the necessity for further research into the role of COX-1 and COX-2 in inflammation.

The unexpected results from COX-1-deficient mice caused Langenbach et al. to state that "Some of the previously physiological roles attributed to COX-1 may not be entirely correct" [41]. The same may hold true for COX-2. At the end of a review of the first COX knockout mice, entitled "Yes, but do they still get headaches?", DeWitt and Smith posed a number of questions that could be quickly and easily answered with the advent of these transgenic animals [34]. How do these animals survive infection? How do they respond to chronic inflammation and is there a modification of their pain response? We would add, how do they resolve an inflammation? To our knowledge, no reports of these studies are yet in the literature. In the rat carrageenin-induced pleurisy we have demonstrated by Western blotting two distinct peaks of COX-2 protein expression (unpublished data). The early peak is associated with production of PGE_2, whilst the later peak, which coincides with resolution of the inflammation, does not produce this prostaglandin. It is possible that this may represent a new function for COX-2 during resolution by producing other prostanoids which may aid in the termination of inflammation. Early results show that treatment with COX-2 selective inhibitors at this later time point of normal resolution, does indeed exacerbate inflammation. Further investigation of these findings is underway.

Deletion of HO genes

In comparison to studies evaluating the role of products of inducible isoforms of the enzymes NOS and COX in inflammation, relatively few studies have addressed the importance of HO-1 in inflammatory disorders. In normal circumstances, HO-1 is undetectable in most cells. The upregulation of HO-1 expression in cells, elicited by a variety of stressful stimuli including heat, metals, UV, inflammatory cytokines (IL-1, IL-2, IL-6, TNFα, TGFβ) and hypoxia, is recognised as a reliable marker of oxidative stress. As such, it is one of several endogenous antioxidant enzyme systems employed by cells to form part of an adaptive response to stress. This includes the activation of superoxide dismutase, catalase and glutathione peroxidase which metabolise free radicals. Oxidative stress is associated with many inflammatory conditions such as atherosclerosis, cerebral ischemia, nephritis and in animal models of acute and chronic inflammation and endotoxaemia. Recently we have demonstrated that HO-1 protein expression is upregulated in the spinal cord of mice with chronic relapsing EAE (A. Tomlinson et al., unpublished data).

The precise anti-inflammatory mode of action of HO-1 is unclear. Bilirubin has potent antioxidant and anticomplement activity, and vasodilator and anti-platelet activity of carbon monoxide may be important, but other stress proteins may also be concomitantly induced. Research from our laboratories indicates that HO-1 is a key endogenous factor in the resolution of inflammation. In models of acute inflammation (complement-mediated, carrageenin-induced pleurisy, antibody-mediated immediate hypersensitivity and cell-mediated metBSA challenged delayed hypersensitivity), HO-1 protein expression and HO activity was maximal and coincident with the suppression and resolution of the inflammatory response [42]. Inhibition of HO activity by tin protoporphyrin, profoundly and dose-dependently increased inflammation. In contrast, the HO-inducer, ferriprotoporphyrin IX suppressed inflammation. Therefore, from this evidence and that provided in the chapter by Willis, one might predict that deletion of the HO-1 gene would result in an inability to resolve acute inflammation.

In a paper by Poss and Tonegawa on the HO-1 deficient mouse, their primary target was the involvement of this inducible enzyme in iron homeostasis [43]. Their observations on inflammation were of secondary interest. HO-1 deficient adult mice developed a serum iron deficiency and pathological iron-loading in tissues indicating the crucial role of this enzyme in iron reutilization. However, these animals developed chronic inflammatory disease by 20 weeks of age, characterised by enlarged spleens and lymph nodes with high $CD4^+:CD8^+$ ratios; neutrophil, macrophage, lymphocyte and plasma cell infiltrates into the liver and occasionally in the lungs; monocyte adherence to vessel walls indicative of vascular injury and glomerulonephritis. The latter may have been a consequence of immune complex deposition or iron overload damage. The same authors addressed how these animals dealt with free radical generation and oxidative stress [44]. Embryonic fibroblasts *in vitro*, from mice deficient in HO-1, enhanced free radical generation by over 200% in comparison to untreated cells when stimulated with hemin. With similar treatment, heterozygous embryonic fibroblasts produced only an 11% increase. Other antioxidants enhanced free radical generation but to a lesser degree. Viability of the embryonic fibroblasts from HO-1 deleted animals treated with oxidants was, however, dramatically curtailed. *In vivo* administration of LPS to model the human disease of septic shock also demonstrated the vulnerability of the HO-1-deficient mice to hepatic injury and mortality [44].

These latter two publications in mutant animals support previous studies which emphasize the importance of HO-1 expression in combating oxidative damage. Whether iron homeostatic mechanisms are the basis for this endogenous protection or only part of the adaptive response requires further research. Animals in which the HO-2 gene has been deleted have been utilised to investigate the role of carbon monoxide in neurotransmission [45, 46]. Oxidative stress in HO-2 knockouts exposed to high levels of oxygen has been examined [47]. Although HO-1 was induced in these animals, the lack of HO-2 increased oxygen toxicity and iron accumulation in lung tissues. These paradoxical results may demonstrate that HO-2 augments the turnover of lung iron during oxidative stress, and that the induction of HO-1 does not compensate. However, the level of HO-1 induction was low in comparison to other systems. Also, HO-2 may perform additional roles which may be necessary for full activation of HO-1 or other protective mechanisms. Recently, Maines et al. have reported the isolation of a cDNA encoding HO-3 [48]. This hemoprotein was a poor heme catalyst but was shown to have enhanced heme binding capacity. It was speculated that this heme-binding capacity may regulate the activity of heme-dependent enzymes including HO-1 (see chapter by Willis in this volume for further discussion of this topic). Discoveries within this field of research are currently outstripping our understanding, and we await future clarification with anticipation.

Interactions between NOS, COX and HO and their products in inflammation

Cirino recently presented a critical review regarding the approach to inflammation and the understanding of multiple factors in inflammation [49]. His approach did not involve an analysis of the inconsistencies of the studies detailing which cytokines and proteins were expressed, enzymes induced and signal transductions activated in myriad systems, but was "to provoke in the reader a critical review of dogmas and current beliefs that most of the time are built on unilateral interpretation of the data". Illustrating this idea with the roles of NOS, COX and phospholipase A_2 in inflammation, it was clear that most published data was conducted *in vitro* on cell lines, and in general was uncomfirmed *ex vivo* or *in vivo*. Interpretation of the importance of a single mediator was often proclaimed, disregarding other mediators regulated by the same stimulus, thus resulting in obfuscation rather than clarity. Our approach has always concurred with that proposed by Cirino. In our laboratories, with regard to the role of inducible enzymes in inflammation, we have always examined the spatial and temporal expression of COX, NOS and HO, their activity, localisation and interactions in a variety of models of acute and chronic inflammation. These parameters have been related to cytokine and growth factor expression, and functional correlates including clinical symptoms, pain perception, gross inflammatory parameters, apoptosis and oncogene expression. A reductionist approach in primary cultures and transformed cell lines has subsequently been adopted for specific interactions, choosing the relevant cell to answer specific questions.

In this chapter we will not attempt an in depth analysis of the inconsistencies of interactions of NOS, COX and HO in *in vitro* and *in vivo* systems, but direct the reader to a number of important papers and reviews. Figure 1 is a somewhat simplistic attempt to give an overview of interactions between NOS, COX, HO and their products, but outlines the main interactions.

Interactions between NOS and COX

Cross talk between NOS, COX, NO and prostaglandins, is extensively documented but investigations have produced conflicting and controversial data. The variability of interactions appears to be dependent upon the *in vitro* or *in vivo* system employed, the levels of enzyme product (particularly NO) and the type and concentration of inhibitors (particularly NOS inhibitors) employed. COX activity can be stimulated or inhibited by NO, and NOS is a similar target for prostaglandins, For a review of the literature see our review [50] and those of Salvemini et al. [51] and Di Rosa et al. [52].

In a series of papers and reviews published by Salvemini and colleagues, a compelling case has been made for the activation of COX enzymes by NO, thus augmenting production of prostaglandins in a variety of *in vitro* and *in vivo* systems

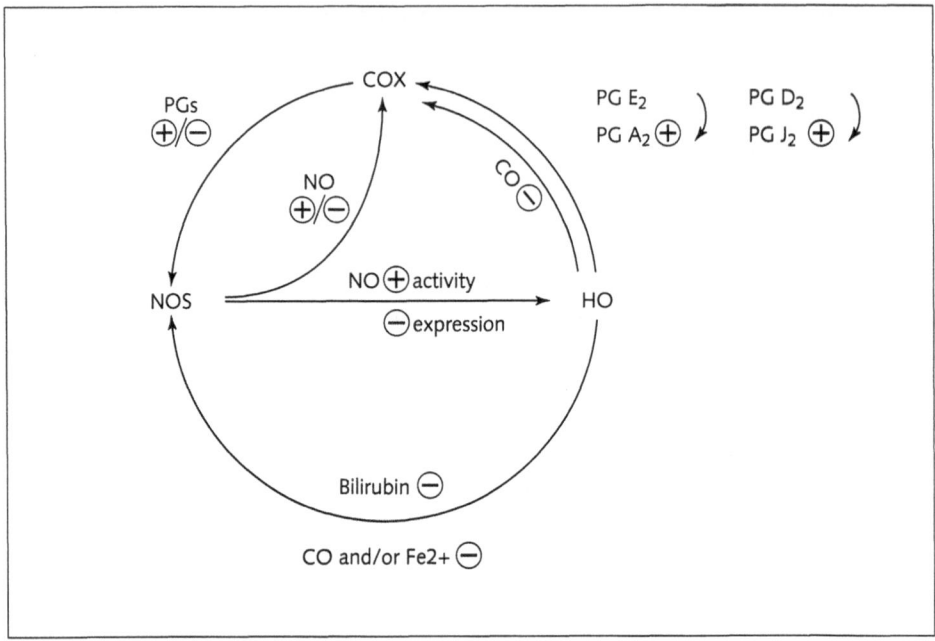

Figure 1
In vitro and in vivo reactions between NOXS, COX and HO.

[53]. Exogenous NO was shown to stimulate the activity of COX-1 and COX-2 in microsomal preparations, pure enzymes and cytokine-stimulated cells by mechanisms independent of the action of NO on soluble guanylate cyclase. Using arginine analogues to inhibit the endogenous production of NO, the prostaglandin release from stimulated RAW 264.7 macrophages was similarly attenuated. The mechanism of action remains to be identified however several are proposed: O_2^- produced during inflammatory events has the ability to inactivate COX, NO by combining with this free radical may limit the availability and thus augment prostaglandin production; NO nitrosylation of cysteine residues in the catalytic domain of COX, by the formation of nitrosothiols can enhance catalytic efficiency; COX can be activated by the hydroxyl radical, which can be formed from the decomposition of peroxynitrite, NO and O_2^- combine to form peroxynitrite. However, the work of Tsai et al. [54] makes it unlikely that NO exerts its stimulatory effects by combining with the ferric heme in COX.

Recent publications on NOS and COX interactions reinforce the variety of previously mentioned possibilities. The inhibition of NOS activity by mercaptoalkylguanidines is also shown to inhibit prostaglandin production in HUVECS and J774 macrophages, with or without stimulation [55]. The mode of action is postulated to be by a direct inhibition of COX-1 and -2, in addition to the effects on iNOS. Interestingly in this latter study, the NOS inhibitors aminoguanidine, L-NAME and L-NMMA did not affect prostaglandin production, except at high doses. Nitric oxide stimulation of COX activity may be important in the regulation of human inflammatory disease. Both exogenous and endogenous NO increased activity of the COX-2 pathway in cytokine-stimulated cultured human airway epithelial cells [56]. Stimulation of NO production by inflammatory mediators in human osteoarthritic cartilage explants similarly stimulated prostaglandin production [57]. However, the endogenous release of NO was found not to modify prostaglandin biosynthesis in a model of zymosan-induced inflammation in the rat air pouch [58]. The findings from this latter publication are in keeping with our observations in the rat carrageenin-induced pleurisy and the murine croton oil granulomatous tissue air pouch. Selective inhibition of COX-2 production by SC 58635 and iNOS inhibition by 1400W in LPS-induced endotoxaemia, also did not provide evidence for NO-stimulated prostaglandin production or prostanoid effect on NO biosynthesis [59]. It is obvious that further investigations are required in this extremely complex area, as NOS and COX are important mediators and modulators in inflammation.

Interactions between COX and HO

Carbon monoxide is a well known inhibitor of cytochrome P450 enzymes including the heme-containing enzyme COX. Depletion of heme levels by HO results in a suppression of cytochrome P450-dependent arachidonic acid metabolism and it is by this pathway that HO most likely exerts its major effect on COX isoforms [60].

Prostaglandins J_2 and $^{12}\Delta$-PGJ$_2$ are metabolites of the prostaglandin PGD$_2$, which is commonly distributed in mammalian tissues. Prostaglandin E$_2$, predominant in mammalian tissues is the precursor of PGA$_2$. Prostaglandins D$_2$, J$_2$, $^{12}\Delta$-J$_2$ and A$_2$, which form a subclass of prostaglandins termed cyclopentenones, are involved in inhibition of cell growth [61] and induction of cellular differentiation [62]. When induced by inflammatory stimuli they are also able to modify the stress response by the activation of heat shock factor-1 [61] , and induction of heat shock protein 70 (HSP 70) [63] and HO-1 [64]. These naturally occuring prostanoids have been shown to exert their effects through the peroxisome proliferator-activated receptor-γ (PPARγ), one of a group of nuclear receptors which are transcription factors [65, 66]. PPARγ activation regulates adipocyte differentiation and glucose homeostasis. In addition, activation of this receptor has recently been shown to inhibit the production of inflammatory cytokines in monocytes and negatively regulate macrophage activation. Thus, metabolites of COX can downregulate the inflammatory response. Research in our laboratories has demonstrated that the effect of the cyclopentenone PPARγ agonists, particularly $^{12}\Delta$-PGJ$_2$, on activated RAW 264.7 murine macrophages was to inhibit the iNOS enzyme pathway [67].

Inhibition was measured as a reduction in nitrite accumulation and iNOS protein expression. This inhibition was correlated with activation of HO-1, whilst other markers of the stress response, HSP 70 and and GRP 78 were unaffected. Thus, the induction of HO-1 may in part account for the anti-inflammatory activity of these COX metabolites. (Evidence for HO-1 inhibition of the NOS pathway is presented in the next section.) Therefore, the generation of eicosanoid metabolites during inflammation may contribute to the resolution of inflammatory lesions through activation of the PPAR system. As early as 1968, work in our department identified the production of $PGF_{2\alpha}$ as an anti-inflammatory agent which coincided with the resolution of the carrageenin-induced pleurisy [68]. It now appears that the cyclopentenones may have assumed that role. Leukotriene B_4 was the first inflammatory mediator demonstrated to activate the PPAR family of nuclear receptors [69]. Activation of PPARα by this lipid mediator was proposed by regulating oxidative degradation of fatty acids, to act as a negative feedback regulator of inflammation. Further investigations of PPAR activation may reveal novel therapeutic targets for the control of inflammation.

Interactions between NOS and HO

The capacity of NO to activate or inhibit metallo-protein-containing enzymes underlies many of its biolgical actions [70]. Nitric oxide interacts with heme-containing enzymes, including soluble guanylate cyclase and NO itself. Heme oxygenase is not a heme-containing protein, however when the substrate heme binds to the enzyme, a transitory hemo-protein is formed [71, 72]. We have shown the capacity of NO to reduce HO activity in tissue homogenates [73]. Accumulating data supports both a positive and negative role for NO regulation of HO expression, nitric oxide donors increased HO-1 mRNA and protein synthesis in aortic vascular smooth muscle cells [74]. Enhanced production of CO however did not negatively feedback to attenuate NO-induced HO-1 induction in this system, despite observation of this mechanism in hypoxia-induced HO-1 expression. Cytokine-stimulated rat aortic smooth muscle cells induced NO, and HO-1 protein was blocked by NOS inhibition suggesting endogenous stimulation of HO-1 by NO [75]. Donors of NO also induced HO-1 activity and protein expression in cultured bovine aortic endothelial cells, although the induction may have been equally attributed to NO interaction with O_2^- and formation of peroxynitrite [76]. Whether the induction or inhibition of HO-1 by NO has any relevance to inflammatory situations requires further examination. Their inter-relationship in hemodynamic regulation in perturbed systems certainly appears important (see chapter by Willis, this volume).

Activation of HO-1 in inflammation may inhibit iNOS activity by the binding and inactivation of the heme moiety by CO [77]. Bilirubin and biliverdin, potent antioxidants may scavenge NO, although it is more likely that they are important in the combination of NO with O_2^- and the formation of peroxynitrite [78]. Activation of nuclear factor κB is required for iNOS production, and elevation of antioxidant molecules may stabilise its inhibitory factor and thus reduce activity [79]. Finally, iron has been shown to regulate NOS activity by controlling nuclear transcription [80].

Conclusion

Choosing the inducible isoforms iNOS, COX-2 and HO-1 as the subject of this book makes an artificial distinction as to their importance in inflammation. They undoubtably produce molecules of extreme importance to inflammation, but to some extent they are self-selecting from the attention already received. However, as we are reminded by other contributors to this book, they are by no means of exclusive interest. The importance of the burgeoning matrix metalloproteinases is examined by Kieseier and Hartung in their chapter on neuroinflammation. Winrow and Blake (this volume) focus on the role of xanthine oxidoreductase in the pathogenesis of rheumatoid arthritis and we have a developing interest in the inducible isoform of tissue transglutaminase. Transglutaminases are a family of calcium-dependent enzymes which catalyse crosslinks between proteins and between proteins and polyamines [81]. To date three isoforms are recognised, plasma transglutaminase (factor XIIIa, an essential clotting factor), keratinocyte transglutaminase and tissue transglutaminase (tTG). A number of observations suggest a role for the latter in inflammation. In response to inflammatory stimuli, tTG accumulates in murine peritoneal macrophages [82], tTG has been shown to increase PLA_2 activity [83] and tTG activity is associated with rheumatoid arthritis [84]. Preliminary results show that inhibition of tTG by cystamine significantly inhibited the inflammation in the rat carrageenin-induced pleural model of acute inflammation and reduced granulomatous tissue in a model of chronic inflammation. All of these, and potentially many other as yet undefined inducible enzyme systems, may provide valid and novel therapeutic targets for resolving inflammatory lesions.

The second artificial constraint in the subject matter of this book is to limit our interest to the inducible isoforms of NOS, COX and HO. The generation of a previously undetectable isoform during an inflammatory event demands attention, but as evidenced here in this chapter, and elsewhere in this book, the constitutively expressed isoforms of NOS and COX have physiological anti-inflammatory roles which can be redefined as pro-inflammatory when affected by early mediators. Evidence presented here demonstrates the anomalies surrounding COX-1, and a role for this isoform in inflammation requires rethinking. Discovery of new isoforms such as HO-3 hint at COX-3 being a reality. Development of hypotheses in which HO-2 and HO-3 may interact for maximal efficiency of HO-1 activity, create new vistas for other enzymes.

A third artificiality is introduced in regarding the exclusive role of any of these enzymes in inflammation. We have attempted to address this topic of enzyme interactions in this chapter and have given a simplified overview of this immensely complex topic. The hemodynamic implications of interaction between NO and CO are a good example of this complexity and the mechanisms require careful dissection.

So what conclusions have we arrived at concerning the pro- or anti-inflammatory role of the various products of NOS, COX and HO in inflammation? The overall conclusion is that temporal and spatial expression, concentration of product and the context in which it is found appear to determine the role. In other words, there are no simple conclusions. Greater certainty will be forthcoming with the development of better tools, specific donors and inhibitors and techniques allowing dissection of complex mechanisms. However, biological systems are best studied as a delicate balance of many factors. Selective inhibition of potentially damaging molecules may be less advantageous than boosting endogenous protective mechanisms. The pages of journals are covered with studies in which interventions are made early into the inflammatory response, and extravagent claims made for their therapy, without examination at the later stages of inflammation. Progress will be made when *in vitro* studies are borne out *in vivo* and regard is given to the adaptive response of the organism.

References

1   Moncada S, Palmer RMJ, Higgs EA (1991) Nitric oxide: Physiology, pathophysiology and pharmacology. *Pharmacol Rev* 43: 109–142
2   Fujii E, Irie K, Uchida Y, Tsukahara F, Muraki T (1994) Possible role of nitric oxide in 5-hydroxytryptamine-induced increase in vascular permeability in mouse skin. *Naunyn-Schmiedebergs Arch Pharmacol* 350: 361–364
3   Kilbourn RG, Jubran A, Gross SS, Griffith OW, Levi R, Adams J, Lodato RF (1990) Reversal of endotoxin-mediated shock by NG-methyl-L-arginine, an inhibitor of nitric oxide. *Biochem Biophys Res Comm* 172: 1132–1138
4   Beckman JS, Koppenol WH (1996) Nitric oxide, superoxide and peroxynitrite: the good, the bad, and ugly. *Am J Physiol* 271: C1424–1437
5   Vane JR and Botting RM (1994) *Biological properties of cyclooxygenase products in lipid mediators.* Academic Press, London 61–97
6   Tomlinson A, Appleton I, Moore AR, Gilroy DW, Mitchell JA, Willoughby DA (1994) Cyclooxygenase and nitric oxide synthase isoforms in rat carrageenin-induced pleurisy. *Br J Pharmacol* 113: 693–698
7   Vane JR, Mitchell JA, Appleton I, Tomlinson A, Bishop Bailey-D, Croxtall J, Willoughby DA (1994) Inducible isoforms of cyclooxygenase and nitric-oxide synthase in inflammation. *Proc Natl Acad Sci USA* 91: 2046–2050

8   Phipps RP, Roper RL, Stein SH (1991) A new view of prostaglandin E regulation of the immune response. *Immunol Today* 12: 349–352

9   Willis D, Moore AR, Frederick R, Willoughby DA (1996) Heme-oxygenase: a novel target for the modulation of the inflammatory response. *Nature Med* 87–90

10   Huang PL, Huang Z, Mashimo H, Bloch KD, Moskowitz MA, Bevan JA, Fishman MC (1995) Hypertension in mice lacking the gene for endothelial nitric oxide synthase. *Nature* 377: 239–242

11   Sheesely EG, Maeda N, Kim HS, Desai M, Krege JH, Laubach VE, Sherman PA, Sessa WC, Smithies O (1996) Elevated blood pressure in mice lacking endothelial nitric oxide synthase. *Proc Natl Acad Sci USA* 93: 13176–13181

12   Huang Z, Huang PL, Ma J, Meng W, Ayata C, Fishman MC, Moskowitz MA (1996) Enlarged infarcts in endothelial nitric oxide synthase knockout mice are attenuated by nitro-L-arginine. *J Cereb Blood Flow Metabol* 16: 981–987

13   Zaharachuk G, Hara H, Huang PL, Fishman MC, Moskowitz MA, Jenkins BG, Rosen BR (1997) Neuronal nitric oxide synthase mutant mice show smaller infarcts and attenuated apparent diffusion coefficient changes in the peri-infarct zone during focal cerebral ischemia. *Magn Reson Med* 37: 170–175

14   Hara H, Ayata C, Huang PL, Waeber C, Ayata G, Fujii M, Moskowitz MA (1997) [3H]L-NG-nitroarginine binding after transient focal ischemia and NMDA-induced excitotoxicity in type I and type III nitric oxide synthase null mice. *J Cereb Blood Flow Metab* 17: 515–526

15   Dawson VL, Kizushi VM, Huang PL, Snyder SH, Dawson TM (1996) Resistance to neurotoxicity in cortical cultures from neuronal nitric oxide synthase-deficient mice. *J Neurosci* 16: 2479–2487

16   Samdani AF, Dawson TM, Dawson VL (1997) Nitric oxide synthase in models of focal ischemia. *Stroke* 28: 1283–1288

17   Iadecola C, Zhang F, Casey R, Nagayama M, Ross ME (1997) Delayed reduction of ischemic brain injury and neurological defects in mice lacking the inducible nitric oxide synthase gene. *J Neurosci* 17: 9157–9164

18   Liew FY, Wei XQ, Proudfoot L (1997) Cytokines and nitric oxide as effector molecules against parasitic infections. *Philos Trans Lond B Biol Sci* 352: 1311–1315

19   Scharton-Kersten TM, Yap G, Magram J, Sher A (1997) Inducible nitric oxide is essential for host control of persistent but not acute infection with the intracellular pathogen Toxaplasma gondii. *J Exp Med* 185: 1261–1273

20   Holscher C, Kohler G, Muller U, Mossmann H, Schaub GA, Brombacher F, (1998) Defective nitric oxide effector functions lead to extreme susceptibility of Trypanosoma cruzi-infected mice deficient in gamma interferon receptor or inducible nitric oxide synthase. *Infect Immun* 66: 1208–1215

21   MacMicking JD, North Rj, LaCourse R, Mudgett JS, Shah SK, Nathan CF (1997) Identification of nitric oxide synthase as a protective locus against tuberculosis. *Proc Natl Acad Sci USA* 94: 5243–5248

22   Yamasaki K, Edington HDJ, McClosky C, Tzeng E, Lizonova A, Kovesdi I, Steed DL,

Billiar TR (1998) Reversal of impaired wound repair in iNOS-deficient mice by topical adenoviral-mediated iNOS gene transfer. *J Clin Invest* 101: 967–971

23 Fenyk-Melody JE, Garrison AE, Brunnert SR, Weidner JR, Shen F, Shelton BA, Mudgett JS (1998) Experimental autoimmune encephalomyelitis is exacerbated in mice lacking the NOS 2 gene. *J Immunol* 160: 2940–2946

24 Gilkeson GS, Mudgett JS, Seldin MF, Ruiz P, Alexander AA, Misukonis MA, Pisetsky DS, Weinberg JB (1997) Clinical and serologicic manifestations of autoimmune disease in MRL-lpr/lpr mice lacking nitric oxide synthase type 2. *J Exp Med* 186: 365–373

25 McCafferty DM, Mudgett JS, Swain MG, Kubes P (1997) Inducible nitric oxide synthase plays a critical role in resolving intestinal inflammation. *Gastroenterology* 112: 1022–1027

26 Mishima S, Xu D, Lu Q, Deitch EA (1997) Bacterial translocation is inhibited in inducible nitric oxide synthase knockout mice after endotoxin challenge but not in a model of bacterial overgrowth. *Arch Surg* 132: 1190–1195

27 Sicher SC, Vasquez MA, Lu CY (1994) Inhibition of macrophage Ia expression by nitric oxide. *J Immunol* 153: 1293–1300

28 Taylor-Robinson AW, Liew FY, Severn A, XU D, McSorley S, Garside P, Padron J, Phillips RS, (1994) Regulation of the immune response by nitric oxide differentially produced by T-helper type-1 and T-helper type-2 cells. *Eur J Immunol* 24: 980–984

29 Casey JJ, Wei XQ, Orr DJ, Gracie JA, Huang FP, Bolton EM, Liew FY, Bradley JA Skin allograft rejection in mice lacking nitric oxide synthase. *Transplantation* 64: 589–593

30 Koglin J, Glysing-Jensen T, Mudgett JS, Russell ME (1997) Exacerbated transplant arteriosclerosis in iNOS-deficient mice: iNOS protects against immune-mediated intimal thickening. *Circulation* 96: 8SS–944

31 Laubach VE, Shesely EG, Smithies O, Sherman PA (1995) Mice lacking inducible nitric oxide synthase are not resistant to lipopolysaccharide-induced death. *Proc Natl Acad Sci USA* 92: 10688–10692

32 Langenbach R, Morham SG, Tiano HF, Loftin CD, Ghanayem BI, Chulada PC, Mahler JF, Lee CA, Goulding EH, Kluckman KD et al (1997) Prostaglandin synthase 1 gene disruption in mice reduces arachidonic-acid-induced inflammation and indomethacin-induced gastric ulceration. *Cell* 83: 483–492

33 Morham SG, Langenbach R, Loftin CD, Tiano HF, Vouloumanos N, Jennette JC, Mahler JF, Kluckman KD, Ledford A, Lee CA et al (1995) Prostaglandin synthase 2 gene disruption causes severe renal pathology in the mouse. *Cell* 83: 473–482

34 De Witt D and Smith WL (1995) Yes, but do they still get headaches? *Cell* 83: 345–348

35 Dinchuk JE, Car BD, Focht RJ, Johnston JJ, Jaffee BD, Covington MB, Contel NR, Eng VM, Collins RJ, Czerniak PM et al (1995) Renal abnormalities and an altered inflammatory response in mice lacking cyclooxygenase II. *Nature* 378: 406–409

36 Gilroy DW, Tomlinson A, Willoughby DA (1998) Inhibition of isoforms of cyclooxygenase in the rat carrageenin-induced pleurisy. *Eur J Pharmacol; in press*

37 Gilroy DW, Tomlinson A, Willoughby DA (1998) Differential effects of inhibition of iso-

forms of cyclooxygenase (COX 1, COX-2) in chronic inflammation. *Inflamm Res* 47: 79–85

38   Editorial „Is there a COX-fight in inflammation?" (1998) *Inflamm Res* 47: 43

39   Wilborn J, DeWitt D, Peters-Golden M (1995) Expression and role of cyclooxygenase isoforms in alveolar and peritoneal macrophages. *Am J Physiol* 268 L294–302

40   Kirtikara K, Morham SG, Raghow R, Laulerderkind SJ, Kanekura T, Goorha S, Ballou LR (1998) Compensatory prostaglandin E_2 biosynthesis in cyclooxygenase 1 or 2 null cells. *J Exp Med* 187: 517–523

41   Langenbach R, Morham SG, Tiano HF, Loftin CD, Ghanayem BI, Chulada PC, Mahler JF, Davis BJ, Lee CA (1997) Disruption of the mouse cyclooxygenase 1 gene Characteristics of the mutant and areas of future study. *Adv Exp Med Biol* 407: 87–92

42   Willoughby DA, Tomlinson A, Gilroy DW, Willis D (1996) Inducible enzymes with special reference to COX 2 in inflammation and apoptosis. In: JR Vane, J. Botting, R. Botting (eds): *Improved non-steroid anti-inflammatory drugs: COX 2 enzyme inhibitors.* Kluwer Academic Publishers, Dordrecht, 67–83

43   Poss KD, Tonegawa S (1997) Heme oxygenase 1 is required for mammalian iron reutilization *Proc Natl Acad Sci USA* 94: 10919–10924

44   Poss KD, Tonegawa S (1997) Reduced stress defence in heme oxygenase 1-deficient cells. *Proc Natl Acad Sci USA* 94: 10925–10930

45   Zakhary R, Poss KD, Jaffrey SR, Ferris CD, Tonegawa S, Snyder SH (1997) Targeted gene deletion of heme oxygenase 2 reveals neural role for carbon monoxide. *Proc Natl Acad Sci USA* 94: 14848–14853

46   Burnett AL, Johns DG, Kriegsfeld LJ, Klein SL, Calvin DC, Demas GE, Schramm LP, Tonegawa S, Nelson RJ, Snyder SH, Poss KD (1998) Ejaculatory abnormalities in mice with targeted disruption of the gene for heme oxygenase 2. *Proc Natl Acad Sci USA* 4: 84–87

47   Dennery PA, Spitz DR, Yang G, Tatarov A, Lee CS, Shegog ML, Poss KD (1998) Oxygen toxicity and iron accumulation in the lungs of mice lacking heme oxygenase 2. *J Clin Invest* 101: 1001–1011

48   McCoubrey WK, Huang TJ, Maines MD (1997) Isolation and characterization of a cDNA from the rat brain that encodes hemoprotein heme oxygenase-3. *FEBS* 247: 725–732

49   Cirino G (1998) Multiple controls in inflammation Extracellular and intracellular phospholipase A2, inducible and constitutive cyclooxygenase and inducible nitric oxide synthase. Biochem Pharmacol 55: 105–111

50   Appleton I, Tomlinson A, Willoughby DA (1997) Induction of cyclooxygenase and nitric oxide synthase in inflammation. *Adv Pharmacol* 35: 27–79

51   Salvemini D, Masferrer JL (1996) Interactions of nitric oxide with cyclooxygenase: *in vitro*, ex vivo and *in vivo* studies. *Methods Enzymol* 269: 12–25

52   Di Rosa M, Ialenti A, Ianaro A, Sautebin L (1996) Interaction between nitric oxide and cyclooxygenase pathways. *Prostaglandins Leukot Essent Fatty Acids* 54: 229–238

53   Salvemini D (1997) Regulation of cyclooxygenase enzymes by nitric oxide. *Cell Mol Life Sci* 53: 576–582

54 Tsai AL, Wei C, Kulmacz RJ (1994) Interaction between nitric oxide and prostaglandin H synthase. *Arch Biochem Biophys* 313: 367–372

55 Zingarelli B, Southan GJ, Gilad E, O'Connor M, Salzman A, Szabo C (1997) The inhibitory effects of mercaptoalkylguanidines on cyclooxygenase activity. *Br J Pharmacol* 120: 357–366

56 Watkins DN, Garlepp MJ, Thompson PJ (1997) Regulation of the inducible cyclooxygenase pathway in human cultured airway epithelial (A549) cells by nitric oxide. *Br J Pharmacol* 121: 1482–1488

57 Manfield L, Jang D, Murrell GA (1996) Nitric oxide enhances cyclooxygenase activity in articular cartilage. *Inflamm Res* 45: 254–258

58 Paya M, Garcia Pastor P, Coloma J, Alcaraz MJ (1997) Nitric oxide synthase and cyclooxygenase pathways in the inflammatory response induced by zymosan in the air pouch. *Br J Pharmacol* 120: 1445–1452

59 Hamilton LC, Tomlinson A Dye LJ, Mitchell JA, Warner TD (1997) Use of selective enzyme inhibitors provide no evidence for cross-talk between inducible nitric oxide synthase and cyclooxygenase 2 in endotoxin treated rats. *British Pharmacological Society Harrogate Winter 1997*

60 Abraham NG, Drummond GS Lutton JD, Kappas A (1996) The biological significance and physiological role of heme oxygenase. *Cell Physiol Biochem 6: 129-168*

61 Amici C, Sistonen L, Santoro MG, Morimoto RI (1992) Antiproliferative prostaglandins activate heat shock transcription factor. *Proc Natl Acad Sci USA* 89: 6227– 6231

62 Kliewer SA, Lenhard JM, Willson TM, Patel I, Morris DC, Lehman JM (1995) A prostaglandin J_2 metabolite binds peroxisome proliferator- activated receptor gamma and promotes adipocyte differentiation. *Cell* 83: 813–819

63 Elia G, Amici C, Rossi A, Santoro MG (1996) Modulation of prostaglandin A1-induced thermotolerance by quecertin in human leukemic cells: role of heat shock protein 70. *Cancer Res* 56: 210–217

64 Koizumi T, Negishi M, Ichikawa A (1992) Induction of heme oxygenase by Δ12-prostaglandin J_2 in porcine aortic endothelial cells. *Prostaglandins* 43: 121–131

65 Lemberger T, Desvergne B, Wahli W (1996) Peroxisome proliferator-activated receptors: a nuclear receptor signaling pathway in lipid physiology. *Ann Rev Cell Dev Biol* 12: 335–363

66 Schoojans K, Staels B, Auwerx J (1996) The peroxisome proliferator-activated receptors (PPARS) and their effects on lipid metabolism and adipocyte differentiation. *Biochim Biophys Acta* 1302: 93–109

67 Colville-Nash PR, Querishi SS, Willis D, Willoughby DA (1998) Inhibition of inducible nitric oxide synthase by peroxisome proliferator-activated receptor agonists: Correlation with the induction of heme oxygenase 1. *J Immunol* 161: 978–984

68 Willoughby DA (1968) Effects of PGF 2-alpha and PGE 1 on vascular permeability. *J Pathol Bacteriol* 96: 381–387

69 Devchand PR, Keller H, Peters JM, Vazquez M, Gonzalez FJ, Wahli W (1996) The PPARα-leukotriene B_4 pathway to inflammation control. *Nature* 384: 39–43

70   Stamler JS, Singel DJ, Localzo J (1992) Biochemistry of nitric oxide and its redox acti-
     vated forms. *Science* 258: 1898–1902
71   Maines MD (1988) Heme oxygenase function, multiplicity, regulatory mechanisms and
     clinical applications. *FASEB J* 2: 2557–2568
72   Henry Y, Lepoivre M, Drapier JC, Ducrocq C, Boucher JL, Guissani A (1993) EPR char-
     acterization of molecular targets for NO in mammalian cells and organelles. *FASEB J* 7:
     1124–1134
73   Willis D, Tomlinson A, Frederick R, Paul-Clark MJ, Willoughby DA (1995) Modulation
     of heme oxygenase activity in rat brain and spleen by inhibitors and donors of nitric
     oxide. *Biochem Biophys Res Comm* 214: 1152–1156
74   Hartsfield CL, Alam J, Cook JL, Choi AM, (1997) Regulation of heme oxygenase-1
     gene expression in vascular smooth muscle cells by nitric oxide. *Am J Physiol* 273:
     L980–988
75   Durante W, Kroll MH, Christodoulides N, Peyton KJ, Schafer AJ (1997) Nitric oxide
     induced heme oxygenase-1 gene expression and carbon monoxide production in vascu-
     lar smooth muscle. *Circ Res* 80: 557–564
76   Foresti R, Clark JE, Green CJ, Motterlini R (1997) Thiol compounds interact with nitric
     oxide regulation of hemeoxygenase induction in endothelial cells : involvement of super-
     oxide and peroxynitrite anions. *J Biol Chem* 272: 18411–18417
77   White KA, Marletta MA (1992) Nitric oxide synthase is a cytochrome P450 type hemo-
     protein. *Biochem* 31: 6627–6631
78   Stocker R, Yamamoto Y, McDonagh AF, Glazer AN, Ames BN (1987) Bilirubin is an
     antioxidant of possible physiological importance. *Science* 235: 1043–1046
79   Xie Q, Kashiwabara Y, Nathan C (1994) Role of transcription factor NF-kappa B/Rel
     in induction of nitric oxide synthase J Biol Chem 269: 4705–
80   Weiss G, Werner-Felmayer G, Werner ER, Grunewald K, Wachter H, Hemtze MW
     (1994) Iron regulates nitric oxide synthase activity by controlling nuclear transcription
     J Exp Med 180: 969–976
81   Folk JE (1980) Transglutaminases. *Ann Rev Biochem* 49: 517–531
82   Leu RW, Herriot MJ, Moore PE, Orr GR, Birckbichler PJ (1982) Enhanced transgluta-
     minase activity associated with macrophage activation. *Exp Cell Res* 141: 191–199
83   Ishitani K, Ogawa S, Suzuki M (1988) Influence of arachidonate metabolism on
     enhancement of transglutaminase activity in mouse peritoneal macrophages. *J Biochem
     (Tokyo)* 104: 397–402
84   Weinberg JB, Pippen AM, Greenberg CS (1991) Extravascular fibrin formation and dis-
     solution in synovial fluids of patients with osteoarthritis and rheumatoid arthritis
     Arthritis Rheum 34: 996–1005

Index

PIR
Progress in Inflammation Research

T Cells in Arthritis

Miossec P.,
Hôpital Edouard Herriot, Lyon, France /
van den Berg W.B.,
University Hospital Nijmegen, Netherlands / **Firestein G.S.,**
UCSD, La Jolla, USA (Ed.)

Rheumatoid arthritis (RA) is the most common and most severe form of inflammatory arthritis. The pathogenesis of RA has been the subject of intense research for several decades. The prevailing hypotheses have changed over the years, and have attempted to incorporate the most recent data. Although T cells represent an important component of the cells which infiltrate the joint synovium, their contribution at a late stage of the disease remains a matter of debate.

The goal of this book is to outline the major arguments and data suggesting that T cells may, or may not, be central players in the pathogenesis of chronic RA. While each of the editors and authors has his/her own bias (as will be clear by reading the respective chapters), our hope is that the readers will enjoy a complete and balanced view of the critical questions and experiments. This is not just an intellectual exercise since the direction of future therapeutic interventions depends heavily on how one interprets the pathogenesis of RA and the contribution of T cells.

Contents

Firestein G.S. and Nguyen K. H.Y.:
T cells as secondary players in rheumatoid arthritis

Fox D. A. and Singer N. G.:
T cell receptor rearrangements in arthritis

Franz J. K., Pap T., Müller-Ladner U., Gay R. E., Burmester G. R., Gay S.:
T cell-independent joint destruction

van den Berg W. B.:
Role of T cells in arthritis: lessons from animal models

Miossec P.
The Th1/ Th2 ccytokine balnce in arthritis

Burger D. and Dayer J.-M.
Interactions between T cell plasma membranes and monocytes

Oppenheimer-Marks N. and Lipsky P. E.
Adhesion molecules in arthritis: Control of T cell migration into the synovium

Bonneville M., Scotet E., Peyrat M.-A., Lim A., David-Ameline J., Houssaint E.:
T cell reactivity to Epstein-Barr virus in rheumatoid arthritis

Sieper J., Braun J.
T cell responses in reactive and lyme arthritis

Breedveld F. C.
T cell directed therapies and biologics

Falta M. T. and Kotzin B. L.
T cells as primary players in rheumatoid arthritis

PIR – Progress in Inflammation Research
Miossec P., et al (Ed.)
T Cells in Arthritis
1998. 238 pages. Hardcover
sFr. 168.– / DM 198.– / öS 1446.–
ISBN 3-7643-5853-X

BioSciences with Birkhäuser

(Prices are subject to change without notice. 9/98)

For orders originating from all over the world except USA and Canada:

Birkhäuser Verlag AG
P.O. Box 133
CH-4010 Basel / Switzerland
Fax: +41 / 61 / 205 07 92
e-mail: orders@birkhauser.ch

For orders originating in the USA and Canada:

Birkhäuser Boston, Inc.
333 Meadowland Parkway
USA-Secaucus, NJ 07094-2491
Fax: +1 / 201 348 4033
e-mail: orders@birkhauser.com

Birkhäuser

PIR
Progress in Inflammation Research

Chemokines and Skin

Kownatzki E. / Norgauer J.,
Albert-Ludwigs-Universität, Freiburg, Germany (Ed.)

The present volume summarizes the state of information on chemokines focussing on skin diseases. The first three chapters deal with the structure and molecular biology of chemokines and their receptors. The following three review information on the interaction of chemokines with lymphocytes, mast cells and eosinophilic granulocytes. One chapter deals with the expression of chemokines in several inflammatory skin diseases. The final chapter reports on in vitro evidence for a growth-promoting activity of chemokines in skin-derived tumor cells.

The volume is of use for the basic scientist interested in practical aspects and for the physician in search for basic mechanisms of skin diseases.

Contents

PIR - Progress in Inflammation Research
E. Kownatzki / J. Norgauer (Ed.)
Chemokines and Skin
1998. 140 pages. Hardcover
sFr. 148.– / DM 178.– / öS 1300.–
ISBN 3-7643-5818-1

BioSciences with Birkhäuser

(Prices are subject to change without notice. 9/98)

For orders originating from all over the world except USA and Canada:

For orders originating in the USA and Canada:

Birkhäuser Verlag AG
P.O. Box 133
CH-4010 Basel / Switzerland
Fax: +41 / 61 / 205 07 92
e-mail: orders@birkhauser.ch

Birkhäuser Boston, Inc.
333 Meadowland Parkway
USA-Secaucus, NJ 07094-2491
Fax: +1 / 201 348 4033
e-mail: orders@birkhauser.com

Birkhäuser

PIR
Progress in Inflammtion Research

Medicinal Fatty Acids in Inflammation

Kremer J.M.,
Albany Medical College, Albany, USA (Ed.)

This volume is a unique assembly of contributions focusing on the biochemical, immunological and clinical benefits of n-3 fatty acids in inflammation.

Leading clinical investigators from fields as diverse as rheumatology, dermatology, nephrology, gastroenterology and neurology have authored chapters. The basic scientific underpinnings of their findings are elucidated as well.

The work is a highly accessible, one-of-a-kind source which will well serve lipid researchers, graduate students, dieticians and members of the food industry.

Contents

PIR – Progress in Inflammtion Research
Kremer J.M. (Ed.)
Medicinal Fatty Acids in Inflammation
1998. 154 pages. Hardcover
sFr. 148.– / DM 178.– / öS 1300.–
ISBN 3-7643-5854-8

BioSciences with Birkhäuser

(Prices are subject to change without notice. 9/98)

For orders originating from all over the world except USA and Canada:

Birkhäuser Verlag AG
P.O. Box 133
CH-4010 Basel / Switzerland
Fax: +41 / 61 / 205 07 92
e-mail: orders@birkhauser.ch

For orders originating in the USA and Canada:

Birkhäuser Boston, Inc.
333 Meadowland Parkway
USA-Secaucus, NJ 07094-2491
Fax: +1 / 201 348 4033
e-mail: orders@birkhauser.com

Birkhäuser